The Painful Truth

Dr Monty Lyman is a doctor and research fellow at the University of Oxford. His first book, *The Remarkable Life of the Skin*, was shortlisted for the 2019 Royal Society Science Book Prize, was a Radio 4 Book of the Week and a *Sunday Times* Book of the Year. He has given many talks at national conferences and has won several essay prizes, including the 2020 Royal Society of Medicine pain essay prize. He lives in Oxford, England.

www.penguin.co.uk

Also by Monty Lyman

The Remarkable Life of the Skin:
An intimate journey across our surface

The Painful Truth

The new science of why we hurt and how we can heal

Monty Lyman

PENGUIN BOOKS

TRANSWORLD PUBLISHERS
Penguin Random House, One Embassy Gardens,
8 Viaduct Gardens, London SW11 7BW
www.penguin.co.uk

Transworld is part of the Penguin Random House group of companies
whose addresses can be found at global.penguinrandomhouse.com

Penguin Random House UK

First published in Great Britain in 2021 by Bantam Press
an imprint of Transworld Publishers
Penguin paperback edition published 2022

A CIP catalogue record for this book
is available from the British Library.

ISBN 9781529176506

Typeset in Caslon 540 LT Std by Jouve (UK), Milton Keynes.
Printed and bound in Great Britain by Clays Ltd, Elcograf S.p.A.

The authorized representative in the EEA is Penguin Random House Ireland,
Morrison Chambers, 32 Nassau Street, Dublin D02 YH68.

Penguin Random House is committed to a sustainable future
for our business, our readers and our planet. This book is made
from Forest Stewardship Council® certified paper.

FSC www.fsc.org MIX Paper from responsible sources FSC® C018179

For Hannah, my wife

Contents

List of Illustrations

Author's Note: Confidence, conflicts and communication

Pain is a universal human experience. It is also a deeply personal one. I believe that one of the best ways of communicating the deep truths of pain is through real stories. Everyone interviewed in the research for this book has given their express permission to be included. For those whose stories I came across many years before conceiving the idea to write about pain, I have applied a double-lock of confidentiality: their name has been changed and the location of meeting moved. If you think you can identify yourself, I promise that it is pure coincidence. All doctors owe a duty of confidence to their patients, a principle that goes back to the Ancient Greeks. The Hippocratic oath states, 'What I may see or hear in the course of treatment or even outside of treatment in regard to the life of patients, which on no account one must spread abroad, I will keep to myself.'[1]

I am not an expert in a specific field of pain, and have no financial incentive to promote a particular treatment. Through research, experience and interviews with pain experts and people-in-pain alike, I have developed opinions on treatments for persistent pain and I express them. But, while I want to communicate the principles of pain, and really hope that my understanding can be of genuine help, my opinions and reflections in this book should not be read as medical advice.

Sticks and stones may break our bones, and words can literally hurt us, so I try my best to avoid the words of war and language that might exacerbate pain. 'Painkillers' will be in most cases termed 'pain relievers'. I must also make an important clarification about the words we use for long-term pain. 'Persistent pain' and 'chronic pain' are the same thing. My preference for the use of 'persistent' over 'chronic' reflects it being a better descriptor for the condition; it is also more readily accepted by individuals and is now becoming widely used. 'Chronic' simply means long-lasting, deriving from the Greek word for time, *chronos*. But it is not common parlance and means different things to different people, with the most dangerous interpretation being 'permanent'. So while I prefer to replace 'chronic' with 'persistent', it must be noted that 'chronic pain' is the most widely used medical terminology. We have to be comfortable using both of these terms interchangeably.

Prologue

'The good news is, there's nothing physically wrong with you . . .'

EVERYTHING WE THINK we know about pain is wrong. That's quite a bold statement. But, by and large, it's true. By 'we', I mean us as a society; I mean most people in and outside the medical establishment. We misunderstand the nature of pain and this misunderstanding is ruining the lives of millions.

And now I – as a freshly minted junior doctor – was witnessing the consequences of this misunderstanding.

It was nine in the evening, and I was coming to the end of an exhausting day shift on the acute medical unit. The sun had long gone down and the ward was bathed in sallow, artificial light. An acute medical unit is a hard place to forget: all the chaos of a low-end department store on Black Friday, soundtracked by a disjointed symphony of beeps and groans. All day, the juniors had been seeing new patients in Accident and Emergency (A&E), who would then be moved into the acute medical unit for further tests and assessments. Later on they would be seen by a consultant, who would make the final decision as to whether the patient needed to stay in hospital or not. Clutching bundles of files in one hand and jotting down barely legible notes with the other, I trailed the on-call consultant as he strode from bed to bed, assessing the day's patients. He was clearly an excellent clinician, albeit a slightly rushed one, and just as I'd finish scribbling down each action plan – *Monitor renal function . . . Bladder scan . . . Organize family discussion* – he would vanish, searching for the next patient on his list.

I dumped the notes and scurried across the blue linoleum floor, trying not to bump into speeding tea trollies or busy nurses. I scanned the next bay, searching for my consultant among the forest of curtains and drip stands. And there he was, already pulling the curtains around our next patient: Paul.

Paul was an IT consultant in his late forties. He lay on the hospital bed with a pillow wedged under the small of his back. He sported a well-worn grimace. Beads of sweat dotted his bald head, occasionally setting off along the ridges of his furrowed brow. Paul had lived with persistent low back pain for the past few years, attributing it to a 'conked-out' office chair. His pain, originally short, sharp twinges confined to a small area of his lower right back, initially came and went. But over the course of the past year, it had become constant and much more intense. Paul had gradually retreated from society: first he gave up on golf, then on seeing friends at the pub, and now he rarely left the house, having taken extended sick leave. His personal life was collapsing in on him as well: his father had died a couple of months previously and last week – apparently unrelated to his pain – his wife had left him. Over the past few days, Paul's pain had spread to the left side of his back and down the side of his right leg. This morning, the pain had been so severe that he couldn't get out of bed. He said he never saw the same GP at his local practice, and felt that they didn't understand his pain; so Paul had bypassed this by getting his son to drive him to the hospital. The A&E doctors had seemed a bit perplexed by elements of his history, so they had erred on the side of caution and ordered an MRI scan. This was to rule out cauda equina syndrome, a rare condition in which the nerves at the base of the spinal cord are compressed.

The scan was completely normal, as was a later, detailed

examination by a neurologist. Blood tests, which can help pick up infectious or autoimmune causes, were also unexciting. My consultant flicked through the notes and explained the findings to Paul: '. . . as you can see, all of your tests are plumb normal. The good news is, there's nothing physically wrong with you . . .'

'So, you're saying it's all in my head?' Paul winced again, so hard that we instinctively winced with him.

'No, of course . . . erm . . . Well, the important thing is that it's nothing serious! We can give you some powerful painkillers to take home, but I think your GP will be best suited to take things from here.'

As we moved away from Paul, the consultant told me to write the most likely diagnoses:

1) Non-specific low back pain
2) Psychogenic pain

'Non-specific low back pain' is fairly self-explanatory: back pain for which no physical cause can be identified. In fact, over 90 per cent of cases of back pain have no identifiable tissue damage.[1,2] 'Pyschogenic' is much more problematic. It implies that the pain is primarily of psychological or emotional origin. Or, as most patients understandably hear it, 'imagined'. Paul went home knowing that there was no serious spinal damage. That is the only positive to come out of this encounter. He was also left not knowing where – or what – his pain was. Either it was caused by ongoing damage in his body that could not be detected – and possibly treated – by medical technology (a terrifying thought) or it was purely psychological: a thought disorder. Paul was left without the dignity of a diagnosis, or even the assurance that what he was experiencing was real. I'm sure that variants of this story happen millions of times a day.

Here's the issue: both of these options are fundamentally wrong. As a society we have fallen prey to what I call 'the painful untruth': the idea that pain is an accurate measure of injury in the body. And following this logic, if it's not in the body, it must be a disorder of the mind. Implicitly or explicitly, most people and most medical organizations are under the thrall of 'dualism': the body and the mind being completely separate entities. This thinking is not just disproven by modern pain science. It's not just inadequate and even offensive to the millions living with persistent pain, who make up roughly one fifth of most populations. This thinking is destroying lives.

This book is an evidence-based exploration of pain that will allow us to see pain – and indeed ourselves – differently. Through stories and studies, it enables the reader to see pain's true nature: pain is a protector, not a detector. Pain is a horrible feeling that urges us to protect our body: we quickly withdraw the vulnerable body part from the suspected source of danger; we reach to guard or support the body part; we avoid a certain behaviour or action. Pain is not a measure of injury. This distinction may seem trivial, but it is utterly transformative. It's a truth that explains how pain can be made in the brain, yet isn't 'all in your head'. It explains the downright weirdness of pain – from the placebo effect to phantom limb pain. It explains why so many people experience pain long after an injury has completely healed. It explains why all pain is real, and that the legitimacy of someone's pain should not depend on the presence of physical damage. Most importantly, it offers an answer to those living with inexplicable pain, and real hope for recovery.

Like most of my peers, I used to see pain as an important – but ultimately uninteresting – symptom of more interesting conditions. How wrong I was. I've also seen that doctors find it hard, in practice, to accept that there is often a very poor relationship between tissue damage and pain. We like things to fall into neat mechanistic or diagnostic boxes; ideally boxes that don't involve patients' emotions or social lives. We like things to be easily measurable, visible and treatable. But pain is messy; it is wonderfully human.

Pain illiteracy has a huge impact on lives and communities.

There is a need to spread the truth. We are living through a pandemic of persistent pain – it is *the* leading cause of disability globally[3] – and we, as a society, are poorly equipped to deal with it. And the prevailing, wrong view of pain doesn't just disadvantage those living with persistent pain; it denies pain relief to those who traditionally have been seen to be 'unreliable' pain communicators: women, ethnic minorities, people with psychological disorders and infants. In groups of people whose conditions haven't enjoyed detailed scientific scrutiny, or whose pain has had no visible or 'measurable' source, doctors have assumed their agony to be exaggerated or invented. Something needs to change.

This is a book for people in pain, people caring for those in pain and people who just want to know more about this fascinating feeling. It is intended to be accessible to individuals of any background. For those who want to dig deeper, see the extensive glossary and many references at the end. I want this book to show people that you do not have to give in to pain, nor spend your life battling it. There is another way. Having said that, this is not a self-help book. I explore a number of evidence-based treatments and summarize them in a non-exhaustive list in the final chapter. But what I want readers to take away is the fundamental principles of what pain is, and to use these as a foundation for anything practical that works.

Where there is pain, there is debate. Pain is emotive in nature, and we all approach the subject with strong opinions. But no human is free from bias. Perhaps the most biasing power of all is our own, personal experience – which cannot necessarily be replicated in others or generalized across the whole population. Recently, I seem to have been cured of longstanding and occasionally severe irritable bowel syndrome (IBS) through hypnotherapy. Hypnosis was something I never heard mentioned at medical school and something I previously sneered at, but my own pain relief has been near miraculous. And while there is good scientific evidence that it works in some people, with some forms of pain, I had to resist my temptation to tout it as a miracle cure for all types of pain, which it certainly is not. Pain is also complex, variable and incredibly difficult to measure. We have mountains of data on it, but these are built on mixed methodologies and conflicting findings. It's no surprise that scientists and practitioners are perpetually locking horns over how to interpret the evidence. And, of course, conflicts of interest run rife, whether it's big pharma or a small clinic: many livelihoods are built on specific understandings of what causes pain and how it can be treated. Having a competing financial interest doesn't make people wrong, of course, but means that another layer of caution should be applied.

Despite conflicting evidence and competing interests, modern pain science has come on by leaps and bounds over the past few decades to reveal an undeniable truth – that pain is a protector – and this truth is forming the foundation of a pain revolution. Understanding this truth ultimately relieves pain. I approached the writing of this book trying to balance healthy scientific scepticism with humility and an open mind. And that's how I hope the book will be read.

1

The Ministry of Defence

Just what *is* pain?

Nothing in life is to be feared, it is only to be understood. Now is the time to understand more, so that we may fear less.

MARIE CURIE

I DON'T LIKE CRICKET. I tell people that I find it boring but, if I'm honest with myself, I don't like the sport because I'm terrible at it. My hand–eye coordination is non-existent and my attention is variable; facts I was regularly reminded of by senior surgeons in my early junior-doctor years. These limitations are not ideal for a sport in which you have to hit a fast-moving ball with a plank of wood to avoid getting hurt (or score 'runs', so I'm told) and catch fast-moving balls to avoid getting hurt (or 'catch opposition players out', so I'm told). Apparently cricket is the second-most popular sport in the world, with around 2.5 billion fans. If you are one of them, please accept my apologies and don't give up on me yet.

I had managed to avoid the sport in my latter years of school, but at the age of twenty-one, on a beach at the tip of west Wales, my five-year cricket-free streak came to an end. A group of friends had booked a cottage on the coast for the Easter weekend. It was a beautiful, sunny afternoon when we arrived, which was all the justification needed for the sporty core of the group to set up a game of cricket. The beach of a small, sheltered cove – roughly one hundred metres in length – was our pitch. My team was 'bowling and fielding', which meant that one of us bowled to individuals in the opposing 'batting' team, and the rest of us took up positions around the beach to either limit the runs of the batting team or – ideally – catch them out. Knowing my lack of cricketing ability well, Tom, our captain, placed me at 'long off', a position so far away from the action that I wouldn't cause much trouble. This was ideal; I could soak in the surroundings with minimal risk of interruption or embarrassment. The tide had just gone out, revealing a beach of wet, caramel-coloured sand littered with smooth pebbles. Imposing cliffs of black mudstone, each topped with a fleece of wild vegetation, flanked either side of the cove. While the day was gloriously sunny, the horizon – miles away across the Irish Sea – was obscured by a green curtain of rain hanging off heavy black clouds. It was beautiful.

THWACK!

I spun around. The game had clearly begun. The ball had been struck by Lyle, the first batsman of the opposing team. When it comes to cricket, Lyle is my opposite: an obstinate, hypercompetitive archetype of the sporty South African. When his family moved to the UK, he towered over his English schoolmates and dominated them at sport, playing cricket for his county and rugby alongside future England internationals. And now this fifteen-stone behemoth had not only launched the plastic beach-cricket ball into the lower stratosphere, but as it made its descent I could begin to see – to my horror – that it was falling in my direction. There was no one else around me. I could either run away or try to catch it. My brain made the executive decision for me not to become a social pariah, so I attempted the latter. The ball was hurtling towards the shoreline, and would land roughly ten metres to my left, so I sprinted to meet it. As the ball was making the last few metres of its descent, I dived for it – every sinew of my under-exercised arms outstretched – and closed my eyes. I hit the ground in a puff of sand. As it settled, an eruption of cheers filled the cove. I opened my eyes. Nestled in my cupped hands was the ball.

Never has a squidgy, luminous-orange lump of plastic seemed so beautiful. Like a gold-panner finding a mammoth nugget, I stood up and held it aloft to a chorus of joy and disbelief. I had caught it. I had caught Lyle out. I had struck down Goliath. *I loved cricket.*

After my twenty seconds of glory were quickly forgotten, I smugly began to jog back across the pebble-studded sand to my fielding position. I felt a fleeting jolt of sharpness in my right foot. My leg retracted slightly but the feeling disappeared as quickly as it came, and I continued running. I'd clearly stepped on a slightly-sharper-than-usual pebble. The game continued, with little further action at my end of the beach. Around ten minutes later, however, something in the corner of my vision caught my attention. Something serpentine was lying on the sand, just behind me, and it seemed to slither whenever I moved. I spun around and took a quick step back, and was relieved to see only a length of inanimate nylon string. But it seemed to be attached to my right foot. Sitting down cross-legged to get a better look, I could see that the sole of my right foot was covered in blood-caked sand. I gently brushed it off to reveal its source. A large, well-rusted fish-hook had embedded itself deep in the arch of my foot. Blood was steadily oozing out of the puncture site. That's when the pain started. The sharp, stabbing waves were reasonably unpleasant – say, a six out of ten – but they seemed to be soothed to a four when my friends gathered round and stared in admiration and disgust. It was as though the social kudos of having this impressive wound reduced its pain. But it was when I retired from the cricket match to sit alone on the small harbour wall, deciding whether to pull it out myself or go to the nearest minor-injuries unit – worrying whether this horrible, corroded object that may well have spent weeks in a fish's mouth might infect me – that the pain went up to an eight. It pushed nine when I simply imagined the process of extracting the hook.

As I spent the best part of the afternoon prising the barbs from my foot, a seed of truth had been planted in my mind. A seed that was at first interesting, and then troubling. The seed was this: pain is really weird. Pain doesn't make sense. When the hook lodged itself in my foot, it caused tissue damage. But the pain I felt fluctuated wildly, despite the issue in the tissue not changing one bit. The pain began when I saw the offending object protruding from my foot, was diminished by the presence of impressed onlookers, grew worse when I was on my own, and even more so when I visualized the angles needed for the hook's exit strategy. The seed in my mind grew into a fundamental truth: pain is clearly not a direct measure of injury. Hurt does not equal harm. We've all experienced this ourselves, whether it's finding unexplained bruises on our legs or the irrational agony of receiving a tiny paper-cut at the end of a stressful day's work. Any hospital's emergency department is a living demonstration of the fragile relationship between pain and tissue damage. There, I found the relationship to be variable even within individual patients: one young man who'd just been stabbed in the centre of his abdomen in a street fight – the knife thankfully missing any organs or major blood vessels – pointed to his open wound and said, 'Isn't that cool, doc?' He wasn't in any pain – that is, until he went out for a smoke and stubbed his right toe on the wheel of a crash trolley. Clutching his toe, but still ignoring his punctured abdomen, he taught me four new expletives in as many seconds.

While the fish-hook in my foot initially didn't cause any pain, there is another true story of a sharp metal object meeting a human foot that shows the opposite extreme of the hurt–harm conundrum. In 1995, a twenty-nine-year-old British builder was clambering down a building's scaffolding and, when he was near enough to the ground, decided to jump down onto a plank of wood. What he didn't realize was that there was a fifteen-centimetre nail protruding up from the plank, which went straight through his left boot. The builder and his nail were brought into hospital; the man was in so much agony that he had to be given the powerful pain reliever fentanyl as well as a sedative. Fentanyl is an opioid – a drug that acts on our body's opioid receptors, producing powerful short-term pain relief. Our body has its own drug cabinet of natural opioids, of which the most famous are probably endorphins, but ever since humans discovered the opium poppy we have been happily hijacking this pain-relief system. Morphine is the best known of these drugs, but fentanyl is one of the most powerful, being about one hundred times stronger. The builder's boot was carefully cut away by the medical team, revealing the nail to have penetrated *between* his toes, causing no injury whatsoever.[1] This phenomenon of severe pain in the absence of tissue damage was recreated in a fascinating lab experiment in the early 1990s.[2] Healthy participants had a dramatic-looking 'head stimulator', resembling an old-school salon hairdryer, placed on top of their heads. The stimulator was wired up to a large controller with an intensity knob. The subjects were told that the electrical current to which they would be exposed usually causes headaches. The trick: there was no electrical current. The whole machine was a sham. Remarkably, half of the subjects felt pain when the stimulator was 'turned on', and reported their pain worsening when the intensity knob was apparently turned up.

Injury is neither necessary nor sufficient for pain. This is a point worth labouring, because most of us are believers in the great 'painful untruth'. In 'most of us', I include most medical professionals. I also include myself, before that corroded fish-hook changed my path.

The Painful Untruth: Pain is a measure of tissue damage

Even if we know, deep down, that pain isn't a measure of tissue damage, many of us act (and many health professionals treat their patients) as though pain is produced in the body and detected by the brain. This isn't neuroscientific nitpicking. Leaving behind this untruth to discover the real nature of pain is the only way we can help those in unnecessary pain. It also leads to a more beautiful understanding of how our brain and body work, and ultimately what it means to be human. Above all, this knowledge is needed to help the fifth of our population living with persistent and often unexplained pain. But, before we look at what pain actually is, we need to understand how we got into this predicament in the first place.

René Descartes, the seventeenth-century French scientist and philosopher, dabbled in religion, mathematics and natural science, creating whole new fields along the way. But he also revolutionized our understanding of pain. At the turn of the seventeenth century, scientists, philosophers and theologians were still divided as to the whereabouts of the source of human thought and feeling. The oldest assumption was that it was the heart – which noticeably changes rate when we feel different emotions. The other school of thought, rooted in the anatomical experiments of the second-century Greek physician Galen, placed it in the brain. This centuries-old debate was clearly one enjoyed by the public, and pithily put by William Shakespeare at the time: 'Tell me, where is fancy bred [. . .] in the heart or in the head?'[3] Descartes firmly placed pain in the brain and the nervous system. Admittedly, he placed the pain (and the seat of the soul) in the brain's pineal gland – which we now know is a sleep-pattern regulator – but we'll allow him that. This was a big step forward. What was also revolutionary was his likening of the brain and nerves to a machine, and introducing the concept of the reflex. In his *Treatise of Man*, posthumously published in 1664, he compared the link between injury and pain to ringing a bell, accompanied by a now-famous drawing of a boy positioning his left foot a little too close to a fire.

Comparing the experience of pain to bell-ringing seems to make sense at face value: our tissue is injured and nerves send up a pain message to the brain via a pain signal. The brain then directly translates this message and voila!: we are conscious of

The 'pain pathway' according to Descartes' *Treatise of Man*, 1664

'If, for example, fire comes near the foot, minute particles of this fire [. . .] have the power to set in motion the spot of skin on the foot which they touch, and by this means of pulling on the delicate thread which is attached to the spot of the skin, they open up at the same instant the pore against which the delicate thread ends, just as pulling on one end of a rope makes to strike at the same instant a bell which hangs at the end.'[4]

pain. The tissue sends the pain signal; the brain reacts. This theory has prevailed – whether explicitly or assumed – for the past four centuries. But it's a theory that is fundamentally wrong. It is the foundation of the painful untruth. If pain were a reflex, a simple signalling system from periphery – the edges of the body, which interact with the outside world – to brain, then we should *always* and *only* feel pain when our tissue is damaged, and the pain should be directly proportional to the extent of injury. But a brief look at the world around us shows that this clearly isn't the case, from the weirdness of my fish-hook encounter or the persistence of pain long after an injury has healed to how modest changes in our mood modify our experience of pain.

Pain has not completely taken leave of its senses, though. Most of the time, short-term pain is a pretty accurate representation of injury. You shut your laptop on your thumb and it hurts; you slam it in a car door and it hurts more! There is clearly a pathway that links the harm of injury to the hurt of pain. Two and a half centuries after Descartes' proposals, the distinguished British neuroscientist Charles Scott Sherrington discovered the pathway. He identified specialized receptors, at the end of nerves, which lay just under the skin's surface and appeared to be activated only by noxious stimuli: things that make us feel pain. He creatively termed these 'nociceptors', from the Latin *nocere* – to harm or hurt.[5] Nociceptors detect damage and danger caused by noxious stimuli, which can be divided into three categories: mechanical (activated when I stepped on the fish-hook), thermal (when Descartes' boy prodded the fire with his toes) and chemical (ranging from stinging nettles to the burning sensation from lactic acid – a by-product of the body's energy generation – during exercise). Specific elements of these stimuli activate nociceptors, triggering a nerve impulse in the direction of the brain. What is particularly fascinating is that there is some crossover: different noxious stimuli can activate the same receptor. Take one of the receptors responsible for detecting noxious-heat, known as TRPV1, for example. This receptor detects temperatures over 43°C, but is also triggered by capsaicin, an active component of chilli peppers. When we eat a chilli, or when it touches our skin, it's not surprising that we feel a burning sensation, as the very same receptors that respond to burning heat are triggered. So, even though there is no change in temperature, we feel heat. This tricks the brain into thinking that we are hot, so the body tries to cool us down by sweating. The capsaicin molecule is soluble in fat, but not in water, so if you take a bite of something slightly over your spice threshold, drinking water is the worst thing you can do: it spreads these noxious molecules far and wide around your mouth, activating more TRPV1 nociceptors. Best to have something fat-based, like yoghurt, milk (which is provided at chilli-tasting events) or – my personal favourite – a mango lassi. Capsaicin is also commonly added to bird feed to deter squirrels and other hungry mammals, which gives us a curious insight into the natural selection of chilli plants. Mammals – at least, those with crunching molar teeth – are likely to destroy chilli seeds, whereas birds are more likely to pass these seeds through their digestive tract undamaged and disperse them over a wide area. Handily, while the capsaicin molecule is interpreted by mammals as hot and painful, it causes no such effect in birds. So, chilli plants have adapted to be appealing only to their feathered friends.[6]

When a nociceptor is activated, it transmits a signal – an electrochemical impulse – along the peripheral sensory nerve in the direction of the spinal cord. This nerve consists of three parts: the cell body (which contains the nerve's DNA and most of its cellular machinery), the dendrite (the branch that leads from the periphery into the cell body) and the axon (the branch that travels away from the cell body). Once the impulse has finished its journey along the peripheral nerve and reaches the spine, it then travels up a nerve in the spine and into the brain. Crucially, though, this is not an uninterrupted highway towards the brain. In another of his many revolutionary discoveries, Sherrington found that the axon of one nerve does not actually touch the dendrite of the next; they are separated by a tiny, microscopic space he called a synapse. Only once the electrical impulse reaches the end of the first nerve in the spinal cord does it cause the release of chemicals called neurotransmitters across this space, which activates the next nerve, which in turn activates the next, and so on, finally transmitting this impulse up the spinal cord and to the brain. Now we're beginning to flesh out what is actually going on in Descartes' boy. When his foot touches the flames, instead of one nerve activating a pain centre in the brain in the way that pulling one

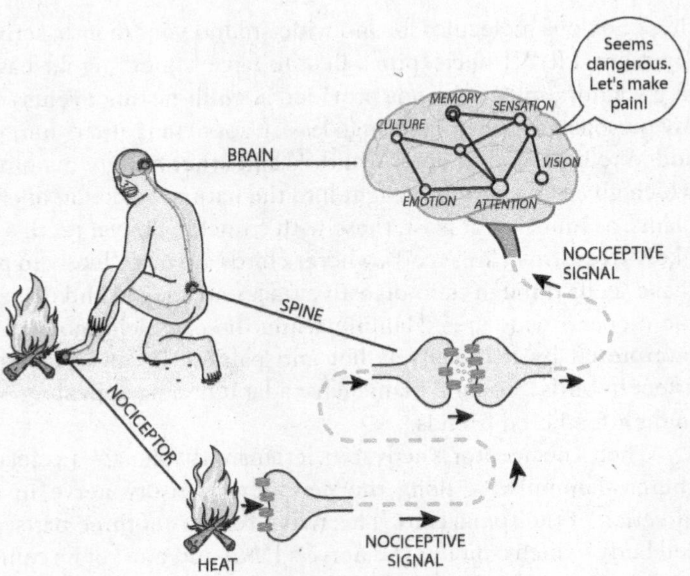

Descartes updated: the nociceptive (danger) pathway

end of a rope rings a bell at the other end, there are actually a number of different nerves that pass the signal to each other before pain is perceived in the brain. It could be tempting – and seemingly logical – to say that we have a 'pain pathway'. Indeed, that is what this transmission process is called in all of my old medical-school textbooks. But here's the truth: there are no 'pain endings', 'pain signals' or 'pain pathways' travelling from the tissues to the brain. Instead, they are nociceptors, and nociceptive signals and pathways respectively. While this is the correct terminology in neuroscience, I am going to call these 'danger receptors' and 'danger signals', because that is what they are. The information these signals convey is that of damage or danger in the tissue, and while they often play an important part in the creation of pain, they are neither necessary nor sufficient for it. Pain is not created in the tissues and does not 'travel up' nerves. Patrick Wall, a British neuroscientist who was perhaps the greatest pain scientist of the twentieth century, and his then PhD student Stephen McMahon (now a world-leading pain expert himself) realized that attempts by scientists and doctors to simplify pain science by using these terms was (and still is) fundamentally wrong. In 1986, they wrote: 'The labelling of nociceptors as pain fibres was not an admirable simplification, but an unfortunate trivialisation. The writers of textbooks will continue to purvey trivialisation under the guise of simplification.'[7]

When the fish-hook was initially lodged in my foot, mechanical danger receptors were sending danger signals along a nerve and into my spinal cord, but something was preventing the creation of pain. The greatest conceptual breakthrough in our understanding of this mystery was proposed by Patrick Wall and Ronald Melzack – an equally revolutionary, Canadian psychologist – in their 1965 paper 'Pain mechanisms: a new theory'.[8] This new theory was the 'gate-control theory'. In essence, they found that there was not a simple one-way path of danger information from peripheral nerves to the spinal cord; there are other, intermediary, neurons that can either let the danger signal through or block its entry, like opening or closing a gate. They proposed that non-painful nerve input (such as touch) could activate inhibitory neurons and stop danger signals from travelling up the spinal cord – 'closing the gate'. This explains why we rub our knee after bashing it on a table. This theory was revolutionary because it offered an explanation for why injury does not necessarily equal pain, and that danger signals can be turned up or turned down.

In keeping with the counterculture of the Swinging Sixties, Melzack and Wall had turned centuries of certainty on its head. Descartes' bell-ringing had ended, and modern pain science had begun. Today's pain scientists are children of this revolution. Over the last decades of the twentieth century, an explosion in pain research took us away from just revealing what pain *isn't*, and we began to see what pain *is*. Pain isn't even decided by the gate-keeping of danger signals allowed through to the spinal cord: pain is created entirely by the brain. In other words, pain is not detected by the brain: it is made by it. And for pain to exist, it requires our conscious awareness. When we are put under general anaesthesia – and take our brain out of the equation – nociception still occurs just as it does when we are awake, but pain itself is never made. (Assuming, of course, that the anaesthetist has done their job properly!) No brain: no pain.

A second important development has been the understanding that there is no single pain centre in the brain. A huge leap forward in recent years has been the ability to image the brain of those in pain. While there are various different neuroimaging techniques, one that will frequently appear in this book is functional magnetic resonance imaging (fMRI), which detects areas of the brain with higher levels of blood flow – and therefore the areas being used – at a specific point in time. Imaging studies show that when someone is in pain, a number of different areas 'light up'.[9] Fascinatingly, the areas activated represent all aspects of what it means to be human: the sensory (detecting what kind of danger signals – if any – are travelling to the brain and where they are coming from), the emotional (such as areas involved with anxiety and stress) and the cognitive (our thoughts, memories, beliefs and expectations). This melange of inputs – unique for each person and for every experience of pain – activates an individualized network of neurons in the brain that creates the perception of pain. This unique network is often termed a 'neurosignature'. When it comes to pain, everything counts. This explains the clunky-yet-accurate name of our most accepted pain theory: the bio-psycho-social model of pain.

It's clear that the 'how' of pain is immensely complex. But it's now clear to science that the 'why' is very simple. And it's genuinely life-changing. It's time for the painful truth.

The Painful Truth: Pain is a protector

'Pain is a protector' is not a definition of pain, but a fundamental truth about it. It is a truth glimpsed in the studies and interviews throughout this book. It is a truth that, when truly understood, explains why pain is so strange and variable, and why pain so often persists after an injury has healed. If we stay as bell-ringers, assuming that pain is a direct measure of injury, these aspects of human existence can never be understood. Knowing that pain is a protector and not necessarily an accurate informant of damage, knowing that pain is trying to help us (even when it becomes overprotective and sometimes life-ruining) is the first step towards healing. Our body has other protective mechanisms that work in concert with pain, such as the immune system, but any definition of pain should be rooted in pain's role as a protector: that is, a feeling that tells us that part of our body is in danger or damaged and that it needs protecting. Whether the body *actually is* in danger, or has been damaged, is another matter entirely. Every definition of pain is a slight compromise, but it needs to be anchored in this core truth. Pain is a horrible feeling that urges us to protect a body part. The International Association for the Study of Pain's definition, updated in July 2020, states that: 'pain is an unpleasant sensory and emotional experience associated with, or resembling that associated with, actual or potential tissue damage'.[10] In the Association's key notes that accompany this definition, it is clear that pain does not equal injury: 'pain and nociception are different phenomena. Pain cannot be inferred solely from activity in sensory neurons. Pain is always a personal experience that is influenced to varying degrees by biological, psychological and social factors'.

We now know that pain is made in the brain. But there's a huge barrier to understanding and explaining this concept. If you were to state this fact to someone living with persistent pain, they would have every reason to ask, 'Are you telling me that my pain is all in my head?' By this we usually mean 'all in the mind', implying that pain can simply be wished away. This could not be further from the truth. Pain is a decision made by the brain – the vast majority of which is outside our conscious control – to tell our conscious mind that we are in danger.

To illustrate this, let's go back to my encounter with the fish-hook and view my pain system in the language of metaphor. Imagine that in my brain is 'the Ministry of Defence': a dispersed network of civil servants, who all work in different departments and sub-departments but whose ultimate goal is the same: protect the body from danger. Some of these workers receive mostly mundane sensory information from the outside world, such as visual input, touch and smell, as well as danger signals. Others are tasked with emotion, past experiences, concentration, core beliefs and future expectations. They are dispersed across the brain but are in constant communication; imagine that they are trapped in a perpetual online meeting. The job of these civil servants is to collectively weigh up the evidence of danger and threat to the body. If there is apparent evidence of danger or damage, the Ministry issues an order to the conscious mind to protect the body; that order is pain.

Let's cut to the action. I've just unexpectedly caught the cricket ball and have bathed in twenty seconds of glory. I'm now gently jogging across the beach, and – unbeknown to me – am about to step on a rusty old fish-hook. Down at the Ministry of Defence, it's looking like a good day . . .

Emotion: I feel amazing! I caught a cricket ball! People love me!

Vision: You've got me to thank for that. I'm the good half of Monty's hand–eye coordination.

Sensation: OK, everyone, sorry to interrupt but I've got a transmission here. I'm getting danger signals from the arch of the right foot. Something small and sharp has scratched the skin.

Vision: Right, let's get the visuals. We're on a beach that's covered in pebbles of different sizes. Nothing particularly dangerous there.

Emotion: Just to let you all know that I'm feeling happy and safe! Peace out!

Memory: Thanks, Emotion. Let me just have a quick look through the archives. Well, Monty's been running on similar beaches in the area every year for the past decade. The only sharp things in the past have been pebbles.

Within a fraction of a second, the Ministry of Defence has to answer one question: is this stimulus dangerous? If it decides that my body is in danger and needs protection, it needs to make my consciousness aware of this fact, and motivate me into action, by making pain. The Ministry's decision is often accurate when it comes to short-term pain, and it has a very experienced workforce, but this decision is only as good as the intelligence the Ministry receives from the outside world. The decision is also powerfully influenced by past experience and future expectation: a Ministry used to regular attacks on the body may become hyper-defensive and trigger-happy. Crucially – and I cannot stress this enough – our conscious mind does not have access to the corridors of power in the Ministry. Pain is its decree: an order that urges our conscious self to protect the body. In my case, however, the Ministry of Defence incorrectly decided that the danger signals came from a harmless source, so pain was not created.

Pain is a conscious translation of our unconscious brain's decision that the body is in danger. V. S. Ramachandran, the eminent Indian–American neuroscientist, puts it well: 'Pain is an opinion on the organism's state of health rather than a mere reflective response to injury.'[11] We can underscore this by suggesting that vision is also the brain's opinion. We believe what we see, and often assume that when light hits our retina our brain simply decodes it, like the camera on a smartphone. Vision appears effortless and it's usually a pretty accurate representation of the world, but optical illusions show that this isn't the case. Have a look at the checker-shadow illusion below.[12]

It's hard to believe, but squares A and B in the left-hand image are exactly the same colour. Compare the squares to the same-coloured bars in the image on the right. Even when we know this fact, we can't change what the brain wants us to see. The brain is adjusting light information so that it makes sense to us, not necessarily to make a completely accurate picture of what is going on out there. It's long been known that much more of the brain is dedicated to generating images than it is to detecting light and colour information – roughly ten times more.[13] Vision is not a measure of light and colour: it is designed to make meaningful sense of objects in the outside world. Pain is very similar: it is not a measure of damage or danger but is instead the brain's unconscious opinion on whether our body is damaged or at risk. Vision is much more than looking; pain is much more than sensing. When we burn our toe or step on a fish-hook, the danger signals that come in from the injury site are clearly important, but they are only as important as what they *mean*.

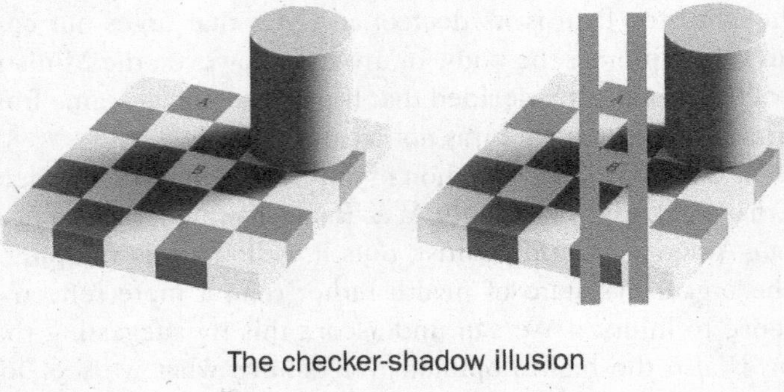

The checker-shadow illusion

It was late spring on the Italian coast a few miles to the south of Rome. But it was 1944. American and British troops had landed on the beaches of Anzio in a daring attempt to outflank the heavily fortified German lines. The Allies had achieved the element of surprise in the amphibious landing, but the Germans rapidly moved all available forces to form a ring around the beachhead. From their vantage points in the surrounding hills, the Germans rained down a maelstrom of artillery fire on the Allies' vulnerable position. It was a bloodbath. In the hospital at Anzio, medical officers were engulfed by wave after wave of wounded. One of these medics was the young Henry Beecher, and what he saw would lead him to become a pioneer of pain medicine. As each wounded man arrived at the hospital, Beecher would ask if they were in pain and whether they wanted any morphine. To his shock, over 70 per cent of the men – some severely injured – didn't report any pain. After the war, Beecher saw the opposite in civilians in Boston wounded by car crashes or industrial accidents: a similar percentage answered 'yes' to these questions. In a paper analysing the two populations, Beecher found that the difference between the groups was not the extent of the injuries, but the meaning behind them, as perceived by the sufferers.[14] On the Anzio beachhead, if a wounded soldier had reached hospital, he knew he was in a place of safety and would probably be shipped home. He was more likely to survive by being injured than not; the chances of survival out on the battlefield were slim. The injured Bostonian civilians, however, were moving from a position of safety to one of danger, so their brains naturally created pain. This is an extreme example – there aren't many instances where being injured is beneficial – but it shows how pain is made by the perception of danger and soothed by safety.

Pain is on our side. Short-term pain (known medically as 'acute pain') is literally life-saving. Those rare individuals who are born without the ability to feel pain live lives of unnoticed and unprotected tissue damage, almost always dying young. There are also people who acquire insensitivity to pain as a result of disease. A leprosy patient I met on a medical trip to East Africa in 2014 could not feel with the tips of his fingers, and repeated unnoticed injuries had left him physically deformed. He told me: 'I would rather feel everything than feel the agony of the shame.' Pain is necessary for life, in every sense of the word. But pain can also be life-ruining; it can eat away at the whole person – their mind, body and social world.

The persistent pain pandemic is widespread and growing, and the medical profession is poorly equipped for it. Persistent pain is also perplexing: there is overwhelming evidence that in most cases the pain persists well after an injury has healed. Although certain cases of persistent pain are caused by a persistent injury in the tissues (usually cancer-related pain or the inflammatory pain of conditions such as gout or active rheumatoid arthritis), and any persistent pain should be checked out with a doctor (who will look for 'red flags'), in the majority of cases pain itself has become the disease. And even if there is persistent tissue damage, it is clear that the extent of injury is poorly related to the amount of pain experienced. In understanding the nature of pain as protector, we can now begin to make sense of another painful truth: that pain has a memory.

Pain remembers

There is a second act to my fish-hook story. In the hours and days following the removal of the offending object, my foot was very sore. The whole arch of my foot was red, inflamed and very tender to touch. Two protective systems were kicking in: my pain system and my immune system. The redness following injury is caused by histamine-containing 'mast cells', the landmines of the immune system, releasing their inflammatory contents. This results in blood-vessel dilation (and redness), widening the roads for the immune army to reach the site of injury. It also makes the skin's danger receptors even more sensitive; just touching the foot caused pain. This phenomenon, in which usually painless stimuli cause pain, is known as allodynia. It's the same process behind sunburned skin feeling particularly sensitive. It reminds us not to mess with the damaged area, and to protect the body part. This is all for our good. For more than a week following the injury, my foot wound oozed pus, suggesting that some nasty infectious agents had travelled on the barbs of the hook. But, after a week or two, my heightened pain and immune systems had died down, and life went back to normal.

Fast-forward to roughly a year later. It was my university summer break and I was on holiday with my parents in the same area of west Wales where I'd injured my foot. The guests of honour were my parents' adorable but hyperactive dogs, Hector and Kiki. As springadors (half springer spaniel, half Labrador), they are energetic at the best of times, but this was the first time they had seen the sea. In an attempt to wear them out, I took them out for a jog along a two-mile-long beach. Suddenly, a sharp, excruciating feeling of pain shot up my right leg. I skipped a step, tripping and collapsing onto the pebble-studded sand, clutching my right foot. I looked at its base. Where the ball of the foot met the arch, not far away from the tiny, faded scar of the fish-hook saga, was the smallest of scratches. I'm not even sure if it bled. It was clearly caused by a sharp stone, but why it caused such an out-of-proportion pain response made no sense at all. That is, unless you know what's going on in the corridors of the Ministry:

Sensation: We've got incoming danger signals from the arch of the right foot. Something small and sharp has scratched the skin.

Vision: OK, here are the visuals: we're on a long beach that's covered in pebbles of different sizes, trying to keep up with some energetic animals.

Memory: I've just been down to the archives. Do you lot remember what happened last time something scratched the arch of Monty's right foot on a pebbly beach in west Wales . . . ?

In an instant, outside of my consciousness, my brain had decided that the scratch represented danger, so it created a good deal of pain to direct my attention to protecting my foot. Ever since these incidents, even though I don't have persistent pain in my right foot, I tend to avoid walking barefoot on pebbly beaches. It's a cycle of hypersensitivity, mild anxiety and avoidance. And it's perfectly common.

My example is trivial compared to the development of life-ruining persistent pain experienced by so many. But the point is that it's the same path. In most cases of persistent pain, our brain has – over time – become overprotective, creating pain even when the damage has gone. This might seem utterly illogical, but it makes sense when we look at the painful truth: pain is a protector. If we pull a muscle in our back, it will almost always completely heal. But, in many cases, in a well-meaning attempt to protect the precious spinal cord, the brain begins to interpret any movement in that area as something potentially threatening, creating pain even though this time there is no damage. History is littered with tales of well-intentioned police forces becoming so hypervigilant after a major crime or a terrorist attack that they end up profiling, attacking and incarcerating innocent people. In a similar way, our brain's Ministry of Defence becomes overprotective, interpreting any innocent muscle movement as dangerous. And the more the brain does this, the better it gets at 'learning' pain, even long after the original tissue damage has healed. In most cases of persistent pain, pain stops being the symptom and becomes the disease. Importantly, this makes the pain no less agonizing, and no less real. But our knowledge of this painful truth is vital – and full of hope.

The springadors who accompanied me on that painful run along the beach are rescue dogs; my parents took them in when they were eight months old. Apart from not being trained at all, it appeared as though they had experienced a happy puppyhood. They were both inquisitive and showered anyone who visited the house with licks and love. That was until the dogs were about two years old and I first invited my friend Josh over. When Kiki, the female of the pair, saw my six-foot-tall, sandy-haired friend walk through the front door, she went berserk. She simultaneously wet herself, started to bark threateningly at Josh and positioned her quivering body between us in an attempt to protect me. She went on to do this every time Josh came over – but only for Josh. Then one day one of my brother's friends, who looks not too dissimilar to Josh, came over, and elicited the same reaction from poor Kiki. Now, it's reasonable to assume that, sometime in Kiki's first eight months of life, a Josh-like man had harmed or scared her or her owner. But now she was threatened by anyone who resembled that old, genuine threat, even though they were harmless. I couldn't get Kiki to be comfortable with Josh by getting angry at her – that would just make things worse. Nor could I sit her down and explain everything logically. What I had to do was slowly provide Kiki with information – evidence – that Josh was safe and friendly: having her in the room while I was spending time with him; getting him to play outside with Hector, her extremely chilled-out brother, and eventually throw the tennis ball for her, too. Slowly and steadily, with a step backwards for every few steps forwards, Kiki began to trust Josh and relax in his presence, and eventually they became great friends.

Whether you prefer to see pain as an overprotective dog or a hypervigilant police force, we need to understand that pain can easily exist in the absence of injury, and overprotection is often the root cause of chronic, persistent pain. The most effective, evidence-based treatments for chronic pain are ones that provide our brain with evidence of safety and reduce the evidence of threat. Trying to 'fight' pain with anger or denial never works, and therapies designed to remove a supposed 'issue in the tissue' rarely work, or rarely work in the way they purport to. Modern medicine loves words of war and fighting metaphors, and perhaps this is appropriate when our body is attacked by foreign invaders (think Covid-19) or by our own cells going rogue (think cancer), but pain is trying to help us. Pain is a friend. Pain is a doctor, teacher and bodyguard. Pain is a guardian angel, rather than an informant of injury. Knowing that pain is *always* trying to protect us, even when it is ruining our lives by being overprotective, is the first step towards living with, reducing and even eliminating persistent pain. Equipped with this simple but revolutionary truth – that pain is a protector – we can begin to enjoy the stories and science behind this fascinating feeling, and even start along a road to recovery.

2

The Painless Five

Insights from the insensitive

I would rather feel everything than feel the agony of the shame.

East African leprosy sufferer

YOU MIGHT ASSUME that it would be great to live without pain, but that depends on whom you ask.

In May 2020 – two months after I had been redeployed to a Covid-19 ward in my hospital – I contracted the disease myself. I noticed my first symptom over dinner. My delicious pre-nightshift vegetable moussaka had absolutely no flavour. A sprinkling of salt just gave each mouthful a salty sensation on the tongue. A generous dusting of pepper made the food distinctly bitter, but again gave it no actual taste. As for dessert, a slice of the chocolate cake I'd enjoyed a few hours earlier was now a lump of disembodied sweetness.

This sudden and profound loss of my sense of smell and taste – a coronavirus-induced anosmia – taught me two important lessons about sensation and perception. The first is that our perception of the outside world (for example, what a food tastes like) is often caused by more than just a single sense or input. In this case, the two primary senses needed for taste are smell and gustation (the sensation of taste on the tongue), while other sensations, such as temperature and sight, are also important for creating the overall perception of flavour. Even sound is involved: bacon has been scientifically shown to taste better when it sounds crunchy.[1]

The second lesson I learned is that sensations we often take for granted are sorely missed when they are absent. Any enjoyment I normally had from eating completely vanished. An alarm system for avoiding rotten and potentially dangerous food disappeared.

In medicine, we learn a lot about sensations and their influence on perception from the stories of those who lack them, whether from birth or as the result of disease or injury. People who cannot feel pain are remarkable not only because they teach us the biological mechanisms behind the experience of pain, and give hope for new pain relievers, but because they are windows into the emotional, psychological and social importance of this complex experience.

Naveed

In his village nestled up in the north of Pakistan, thirteen-year-old Naveed was something of a celebrity. Crowds of curious shoppers would throng around the trail of hot coals laid out at the side of the street as he nonchalantly walked barefoot over them. Once sufficiently remunerated by the awestruck onlookers, the boy would then theatrically brandish a knife. Without so much as a wince, he would slowly drive the blade into his arms. This boy had seemingly gained complete mastery over what should be indescribable pain. Naveed's fame began to spread.

His fame happened to spread just as Geoff Woods, a Cambridge geneticist, was researching neurological diseases in a nearby region of Pakistan. When Woods was asked by local doctors to assess Naveed, he knew that he couldn't miss a chance to see the boy who felt no pain. The condition, known as congenital analgesia (also known as congenital insensitivity to pain), is exceedingly rare. In 1932, the American neuropsychiatrist George Van Ness Dearborn had documented his findings from assessing a mystifying man. This middle-aged gentleman was to all intents and purposes a distinctly average individual, apart from the fact that he had managed to sustain a number of injuries (including accidentally shooting off his left index finger with a pistol) without experiencing any pain. Dearborn admitted that 'in short, we know as yet far too little about the nervous system to warrant a single guess as to the neuropathology of such a case as this'.[2] In the seventy years between Dearborn's case report and Woods receiving news of the painless boy in Pakistan, little had changed: knowledge was still scant. Meeting Naveed would be an incredible opportunity, not to see a medical curiosity but to look through a living window into the mechanisms of pain and – potentially – its treatment.

Around the time of Naveed's fourteenth birthday – just as Professor Woods was preparing to set out to meet him – the boy was showing off to his friends by jumping off the high roof of his family home. After hitting the ground, Naveed stood up, dusted himself off and walked away, seemingly unscathed. Soon afterwards, and with little warning, he rapidly lost consciousness and died. It later turned out that he'd hit his head so hard in the fall that it had caused a catastrophic bleed in his brain, but he hadn't felt even a bump.

Members of Geoff Woods' team at Cambridge and a group of doctors in Pakistan contacted Naveed's family. It transpired that numerous members of this family – as well as two other closely related families within the same clan in northern Pakistan – also had congenital analgesia. Many had scars and broken bones, and had bitten off the front portions of their tongues. Fascinatingly, although they felt no pain at all, their senses of touch, pressure and temperature were preserved.

On the surface it seems as though a literally pain-free life would be a blessing. But the testimonies of those with congenital analgesia suggest otherwise. I spoke to a doctor who had once treated the injuries of a young British boy of Pakistani heritage (maybe he was a distant relative of Naveed). 'I remember that, although he didn't feel any pain, he seemed chronically downcast, depressed,' the doctor recalled. 'He'd spent most of his young life constantly checking his body to make sure he hadn't stepped on glass or burned his hand. You know that paranoia after a dentist tells you not to drink anything hot or eat anything for a few hours after anaesthetizing your mouth? Imagine having that for your entire life. Miserable.'

Instead of living a life of physical bliss, people with congenital analgesia spend their days looking over their shoulder, trapped by a fear of undetected tissue damage. These two cases clearly show that short-term pain is good for us; it is necessary for survival. Pain is life-saving. We are soft, and lots of things out there in the world are hard, sharp, hot and sometimes carnivorous. Pain is both an alarm system and a teacher, altering our behaviour so we can protect ourselves from injury and improve our chances of survival. Sufferers of congenital analgesia have to learn this through other senses: the sight of bleeding, the smell of burning flesh, the popping sound of a bone breaking. But even this is ultimately inadequate. Not knowing that your appendix has burst, for example, can be lethal. In Woods' study, none of Naveed's affected relatives had survived past their teens.

Brain scans and nerve biopsies of members of this family from Pakistan found no structural or anatomical abnormalities. But, when their DNA was analysed by Woods' team in Cambridge, it was clear they all shared a single mutation in the SCN9A gene.[3] This gene encodes a sodium channel called $Na_v1.7$, which is found in nerves that supply the skin and internal organs. Sodium channels are, in essence, gateways at the end of nociceptors, which can activate nerves by allowing an influx of sodium into them. Sodium, which is positively charged, enters the negatively charged interior of the nerve, and this rapid flip in electrical charge is what sets a nerve impulse in motion. $Na_v1.7$'s unromantic name actually explains its function quite well. 'Na' is the atomic symbol for sodium, which travels through this channel; 'v' stands for the change in voltage across the nerve's membrane; and 1.7 simply means that it is the seventh of these channels to be discovered. Importantly, it is almost exclusively found in nociceptors, which activate nerves that project danger signals from the tissues, potentially causing pain. When a dangerous stimulus is detected, a 'generator potential' – a small electrical impulse – travels along the nerve and reaches the $Na_v1.7$ channel, where it is amplified into a larger signal and sent along the nerve. This impulse is then passed onto nerves in the spinal cord and on, into the brain, where it might finally be interpreted as pain.

Research at the University of Oxford in 2019 demonstrated that $Na_v1.7$ is essentially a volume-control knob, and that, for some people, mutations affecting this sodium channel mean that nociception is constantly dialled down to zero.[4] That's why even when Naveed's feet were being burned and damaged by the hot

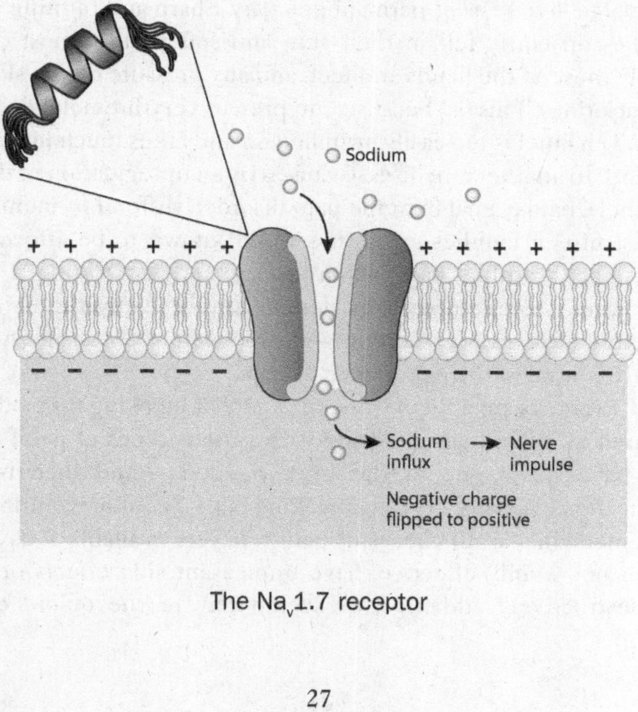

The $Na_v1.7$ receptor

coals, his peripheral danger-detecting nerves were silent. In 2015 a team from University College London revealed that the $Na_v1.7$ gene mutation also reduces pain through another, fascinatingly different mechanism. They found that in mouse models and in those rare humans with the $Na_v1.7$ deletion, levels of an opioid found naturally within our bodies, called enkephalin, were significantly increased.[5] Remarkably, one individual was able to feel pain for the first time in his life after being given the widely available opioid blocker naloxone. Perhaps the value of pain is most evident in the fact that treating pain-insensitive individuals requires making them feel it.

But there is another side to the coin. With $Na_v1.7$ being the volume knob that sets the gain on our pain sensitivity, other mutations can have the opposite effect to Naveed's condition: turning pain up to the max. Sufferers of the rare disease primary erythromelalgia live lives of permanent agony. Sharp and burning pain is near-constantly felt in their skin and soft tissues, most commonly those of the hands and feet, and any pressure on the skin is excruciating. This is because, in primary erythromelalgia, the $Na_v1.7$ channel is too easily switched on and takes much longer to turn off. In another rare disease caused by an upregulation of these channels, paroxysmal extreme pain disorder, sufferers – members of just fifteen families across the world known to be affected – experience terrible pain when eating and defecating.[6] This unassuming receptor has been demonstrated – sadly, through the suffering of those for whom it works too well or not at all – to be utterly central to the experience of pain.

If scientists are able to work out a way of blocking this sodium channel in individuals suffering with various types of pain, they will have made one of the most needed – and lucrative – medications in history. A drug blocking $Na_v1.7$ could revolutionize pain medicine, as all classes of pain relievers available today are either not wholly effective, have unpleasant side effects or can be destructively addictive, as seen today in the opioid crisis devastating communities across the United States. Targeting $Na_v1.7$ makes a lot of sense, as there is very neat science showing its central role in danger detection – in the handful of humans in the world with non-functioning $Na_v1.7$ channels, who do not experience pain at all, and the few even more unfortunate humans who have an excess of these channels and live in constant pain. $Na_v1.7$ is also absent in heart-muscle and brain cells, avoiding the risk of side effects in these organs, ranging from dizziness to fatal heart arrhythmias.

Unsurprisingly, following the discovery of $Na_v1.7$ fifteen years ago, every large pharmaceutical company has set out in search of this holy grail of analgesia. But these years have been littered with research failures, showcasing how unfathomably difficult it is to create a new drug, even with great science behind it. The central conundrum is the need for the drug to be super-selective. $Na_v1.7$ is a member of a family of nine receptors with very similar structures but differing roles across the body, and a messy drug that blocks all of these would open a Pandora's box of side effects. Drugs that block all of these receptors are already in use – including the local anaesthetic lidocaine – but although they are perfect for completely numbing a tiny area of the body before, for example, the excision of a skin cancer, they would be both useless and dangerous if used to treat whole-body or persistent pain. To specifically block $Na_v1.7$, a tiny, unique area of the receptor needs to be identified and a similarly tiny object needs to be created to block it.

In an effort to find a suitable small molecule, the big pharma companies decided to consult some individuals who knock out nerve receptors for a living: tarantulas. In 2018, the company Amgen identified a peptide (essentially a small protein) in the venom of the terrifyingly named Chinese earth tiger tarantula that selectively blocks $Na_v1.7$.[7] But it will be some years, and a few tens of millions of dollars will be spent, before we see whether it has any clinical effect on chronic pain in humans. Even if it works, this new wonder-drug would have to be dosed incredibly carefully in order to enable protective pain. It's important to remember that people like Naveed, with no $Na_v1.7$ function at all, would have desperately loved to feel pain. Perhaps a solution is for a $Na_v1.7$ blocker to be combined with other painkillers, such as opiates, which would be used in lower doses to reduce unwanted side effects.

The $Na_v1.7$ receptor is a gateway between the outside world and our inner being. Rare, remarkable individuals like Naveed have shown that if this gateway is locked shut or left permanently open, the consequences can be devastating. $Na_v1.7$ has drastically improved our understanding of nociception over the last two decades. However, the continued discovery of pain-free people is clearly demonstrating that $Na_v1.7$ is certainly not the only danger pathway that leads to pain. Individuals with congenital analgesia identified from four families worldwide have mutations in another sodium channel, $Na_v1.9$.[8] Six members of the Marsili family from Tuscany, famed locally for their hardiness and apparent high pain thresholds, have recently been found to harbour a rare – perhaps unique – mutation in the ZFHX2 gene.[9] This gene is not responsible for making a specific receptor like $Na_v1.7$, but, rather, regulates a number of genes involved in other aspects of pain, which could hopefully provide tantalizing targets for new pain relievers.

Naveed and other sufferers of congenital analgesia demonstrate the crucial role of pain in human survival. The experiments of their lives were observed and the case seemed closed: life without pain is miserable and short. In 2019, however, someone appeared on the scene, and she defied all the rules.

Jo

'I honestly don't know what pain is,' the Scottish septuagenarian told me.

It's always interesting meeting a patient with a rare disease, but it's not every day that I get to interview the only known sufferer in the world. Although I very much doubt she would ever call herself a sufferer, nor a patient. Like Naveed, Jo has a genetic mutation that means she never feels pain. But the mutation has a very different effect on her life.

'I have such a happy disposition that I'm positively irritating sometimes!' she chirped. Jo's energy was infectious.

Not only does Jo not feel pain, but she also doesn't feel anxiety or fear. None whatsoever. She felt no rush of adrenaline when her car veered off the road and overturned, nor when she was left dangling over a ravine on a zip wire in Montenegro a week before our interview. 'I guess the only downside to the mutation is that it affects my short-term memory. I'm always forgetting where I leave my keys. But, then again, that doesn't bother me very much, either.'

Jo has a mutation in a pseudogene (a largely inactive copy of another gene) called FAAH-OUT. This was long assumed to be a piece of 'junk' DNA, a piece of our genetic code that doesn't actually code for anything. But it turns out that it plays a crucial role in regulating the fatty acid amide hydrolase, or FAAH, gene. The enzyme this gene produces, FAAH, normally breaks down anandamide, an endocannabinoid that is produced by the body and binds to cells' cannabinoid receptors, which play roles in mood, memory and pain. The most famous cannabinoid is tetrahydrocannabinol (THC), the most potent psychoactive ingredient in cannabis. Indeed, 'ananda' is the Sanskrit word for bliss. In Jo's case, a lack of FAAH means that her anandamide is not broken down. To some extent, Jo is permanently high.

'I think the scientists who named the FAAH-OUT gene were having a bit of a laugh, as the high levels of cannabinoids in my blood make me relaxed, carefree and a bit forgetful: essentially, I'm a lifelong stoner. Far out, dude.'

What is just as remarkable as her condition is the way in which it was discovered. Jo didn't realize she was any different until the age of sixty-five. Over a four-year period, her hip would occasionally give way when she walked, although doctors thought little of it as she felt no pain and it hardly bothered her. Eventually an X-ray revealed that the joint was severely damaged by osteoarthritis, a disease normally accompanied by agonizing pain. When she had her hip replaced, the surgeons noted that – unbeknown to her – both of her thumbs were also grossly damaged by osteoarthritis. Her anaesthetist in Inverness was already flummoxed by the Scot's incredible pain threshold, but when Jo felt nothing following notoriously painful thumb surgery (a trapeziectomy, which involves removing a small bone in the wrist) he contacted James Cox, a geneticist at University College London. Cox had long been studying remarkable painless individuals; he was Geoff Woods' postdoctoral student in Cambridge when Naveed's family in Pakistan were first investigated.

'After I was told about the mutation, in hindsight things started to make sense,' recalled Jo. 'As a child, I didn't know that I'd broken my arm until the bone started healing at a strange angle, and I only realize that my hand has been on the cooker too long when I smell burning flesh. It helps that I'm vegan.' My mind was just beginning to process this unpleasant image when Jo nonchalantly continued: 'Actually, right now there's a square-shaped bruise on the top of my foot. I have no idea how it got there. I must have dropped something on it'.

What I couldn't understand was how she had lived into her seventies, especially without losing any limbs, if she doesn't feel any pain at all. The secret to Jo's survival, compared to others who have congenital insensitivity to pain, may be due to another mysterious quirk of the FAAH-OUT mutation. The many wounds she sustained throughout her life healed remarkably quickly, often without scarring. Mouse models with mutations in the FAAH gene also support accelerated skin-wound healing.[10] This could be because one of the fatty acids usually broken down by FAAH stimulates the proliferation of skin cells.

Jo is not just an oddity. She is a miracle. Pain and fear are our teachers and guides through this dangerous life, long assumed to be necessary for survival. Jo has neither of these. It also seems to make sense that a life without pain would be a bit dull, with no pain to motivate us or make pleasure rewarding. But Jo's long life appears to be a perennially satisfied one. Throughout 2021 she is undergoing experiments intended to make her feel pain for the first time, and she's looking forward even to that. This lack of concern is one reason why Jo is so different from Naveed and other individuals with congenital analgesia, none of whom have demonstrated significantly low anxiety levels. This could well be due to increased signalling at CB1 receptors, a type of cannabinoid receptor that is activated by both internally made anandamide and cannabis-derived tetrahydrocannabinol. Increased activation of this receptor reduces anxiety and helps the mind and body cope with stressful situations.[11] Jo is a living, talking demonstration of how pain and mood are intertwined.

Naveed's and Jo's single mutations reveal that pain is controlled and modulated in very different ways. Naveed's SCN9A mutation blocks pain at the gateways of the nerves, whereas Jo's FAAH-OUT mutation increases the amount of endogenous painkiller in the body. Jo's confirmation of the analgesic power of anandamide provides hope for new painkillers working with the endocannabinoid system. It is tempting to wonder whether these could also recreate Jo's happy-go-lucky disposition, treating anxiety and depression as well.

Sadly, we have known about FAAH for some time and not one of the many FAAH inhibitors developed have progressed to become medications. And, tragically, in this race to find the miracle painkiller, some have paid the ultimate price. In a 2016 clinical trial in France, a new 'FAAH-inhibitor' drug had numerous off-target effects on the human nervous system, leading to the death of one study participant and permanent brain damage in four others.[12] Blocking an enzyme that breaks down different substances in different tissues of the body is fiendishly difficult. But hopefully the discovery of incredible people like Jo will be a catalyst for new research into treatments for persistent pain. Thanks to the explosion of public interest in her case in 2019 – including her appearance on national television eating Scotch bonnet chillies – more and more individuals who claim to never experience pain are coming out of the woodwork, volunteering to be lab rats. Perhaps insensitivity to pain is more common than previously thought.

Candice

'A bowling ball landing on my spine.' 'Like trying to put your lower lip over your entire head.' 'The ripping apart of your pelvis like tectonic plates.' If you ask Google – or anyone who has given birth – what childbirth feels like, prepare to be both entertained and terrified. After delivering my first baby as a medical student (and by delivering, I mean letting an unflappable midwife direct my passive, shaking hands as I stared at the scene with a mixture of horror and awe), I have the deepest respect for those who have gone through labour. Childbirth is a heroic act of grit, stamina and bravery. While no two labours are the same, most women experience pain at some point during childbirth. That is, unless you are Candice.

'For my first birth I was really nervous and thought it was going to hurt like hell,' the teaching assistant from Northamptonshire told me, recalling when her first labour started. 'I was Christmas shopping at Tesco one day, when I felt a sharp sensation across my abdomen. It was strange but not really painful.' She thought that this was a false labour, or Braxton Hicks, contraction, as it was a week before her due date and she'd experienced a few of these false contractions earlier on in her pregnancy. Candice finished her shopping and came home. It just so happened that a friend she'd met at antenatal classes then popped over for a coffee and, due to this woman's extreme anxiety about giving birth, had brought a community midwife along with her. During coffee and cake, Candice felt another odd tightening of her abdomen: 'I think I'm having contractions,' she told them. The midwife wasn't at all convinced that the relaxed and chatty Candice was going into labour, but suggested she go to the nearby hospital anyway. When Candice arrived at the hospital her cervix was only three centimetres dilated, but after about ten minutes she had dilated to ten centimetres. What happened next was so unremarkable it is utterly remarkable: 'I had about two contractions and then I just pushed my son out. He popped out like a rugby ball.'

Candice knows what it's like to feel pain, but her suspicions that she had a higher pain threshold than most of her friends were confirmed after she had required no pain relief whatsoever for any of her three labours. Members of Professor Wood's laboratory, at Cambridge, were intrigued by painless labour, and in 2020 revealed that this is likely caused by a 'natural epidural' gene. Dr Michael Lee and other researchers identified around 1,000 women across the UK who had not needed any pain relief during their first labour.[13] The women were then subjected to a number of types of experimental pain: a thermode to the forearm gave both burning and freezing sensations, and mechanical pressure was applied with a blood-pressure cuff. Perhaps unsurprisingly, these women had higher pain thresholds than a control group of women, and, importantly, the researchers found no differing cognitive or emotional abilities between the groups that could explain the difference in pain thresholds. Note that by 'pain threshold' we mean the minimum intensity at which someone perceives a stimulus to be painful, compared with 'pain tolerance', which is the maximum amount of pain one can bear.

The team then sequenced the genomes of both groups, and found that in the painless-labour group there was a much higher prevalence of a mutation in the KCNG4 gene. This gene encodes a channel – $K_v6.4$ – that acts as a gateway at the end of nociceptors, controlling the flow of potassium in nerve cells in a similar way to how $Na_v1.7$ does with sodium. This mutation is crucially different from Naveed's, however, in two main ways. First, it does not render the gateway completely useless, so these patients can still feel pain. The mutation instead makes the channels defective, so that they require a much greater stimulus (i.e. strong contractions during labour) to send a nociceptive impulse from the periphery to the brain and for the patient to feel pain. This is perhaps the most literal way in which someone's pain threshold can be raised. Second, these mutations are much, much more common than Naveed's – or Jo's – insensitivity to pain: roughly one in a hundred women carry this genetic variant. The discovery of this variant isn't just a medical curiosity: if a novel painkiller that specifically targets the $K_v6.4$ channel is developed, it could present a revolutionary way to provide labour-pain relief with no side effects for the mother or baby – something we do not yet have.

Candice also shows how genetic variations influence how each one of us feels pain. There are likely many genetic variations that influence pain thresholds and pain experience that we do not yet know about. At my school, during the inescapable playground fights, the boy to avoid was Duncan, a large, red-haired Scot. As well as being at least a foot taller than the rest of us, he claimed that his terrifying strength lay in the fact that 'Scottish people don't feel pain!' He certainly packed a powerful punch. Fascinatingly, studies have shown that red-haired people may indeed be more resistant to a number of types of pain, including electrical pain, but are more sensitive to thermal pain.[14] This could well be due to little-explored effects of a mutation in the melanocortin 1 receptor (MC1R) gene, which produces the red hair pigment.

Peter

'I've got a front of brass, but feet of clay!'

Peter was certainly one of the most entertaining patients on the vascular surgery ward. Even on the morning of his right-foot amputation, the retired English professor was able to cheerfully quote Byron to the team of surgeons on the ward-round. Peter had a portly frame, ruddy cheeks and a wild net of greasy grey hair that clung to the back of an otherwise bald head. As one of the junior doctors on the ward, I had got to know him well, particularly as he – clearly bored rigid – liked to beckon me over at various points in the day to teach me a famous quip or line of poetry. I knew his story well – the chain of events that led to his foot having to be amputated. It was an all-too-familiar tale of a painless path to destruction.

Peter had been diagnosed with type 2 diabetes ten years previously at a routine check-up with his GP. He'd been put on a tablet to reduce his sugar levels but he freely admitted a rather laissez-faire approach to taking it. Six uneventful years passed, until one summer afternoon he noticed a trail of blood following him across the beige carpet of his conservatory. It was dribbling out of the ball of his right foot, which, on closer inspection, had been punctured by an inch-long thorn, most of which still rested deep within the flesh. Peter had clearly stepped on it while out gardening barefoot but, curiously, had not noticed any pain when the thorn entered, remained embedded or was removed from his foot. The wound seemed to heal over quickly and he got back to his retirement duties of gardening and countryside rambling.

Peter was developing diabetic neuropathy: diabetes-induced nerve damage. His high levels of blood glucose, reduced insulin-signalling and dysregulated levels of cholesterol were slowly causing inflammation to the nerves in his feet and to the blood vessels that supply them with nutrients. The first nerves to die off completely were probably his C-fibres, which carry slower waves of nociception, as these nerves are not protected by the insulation of myelin – a fatty substance that surrounds most nerves. Then other nerves began to become damaged and die, progressing from his toes up his lower limbs until both of his feet were completely insensitive to pain.

As he continued to inspire his poetry-writing with long rambles in the Oxfordshire countryside, Peter slowly began to sustain damage to the bones and joints within both of his feet, but particularly his left foot. Usually, an injury to a limb causes an abnormally increased sensitivity to pain in the affected area, our nerves shouting at our brain to give the limb some time to rest. In many with diabetic neuropathy, however, the dead nerves have no signal to send.[15] Peter kept rambling while the bones in his foot began to wear away, the continuous movement and chronic inflammation weakening the bones' structure, even causing them to break. All of this was completely unbeknown to Peter until his left foot was so swollen he could no longer fit it into his hiking boot. A closer inspection revealed that the foot wasn't just enlarged, but that the arch of his foot had completely collapsed. When Peter visited his doctor's surgery – almost a decade since his last appointment – his GP explained that this process of painless, steady damage to his left foot is known as 'Charcot foot', named after the French neurologist Jean-Martin Charcot. This could potentially be salvaged with surgery.

Peter's GP sensibly had a glance at his other foot, the right one, too. Around the ball and arch of this foot, a three-centimetre-wide ulcer bore deep into the skin. Bone was visible. This severe diabetic foot ulcer could have started from any small injury – perhaps even the garden thorn a few years earlier – going completely unnoticed in Peter's painless peripheries. Diabetic foot ulcers may sound trivial, but they often have terrible consequences. A 2017 study at the University of Leeds recruited patients who had been diagnosed with an infected diabetic foot ulcer and followed them over the course of a year. At the end of that year, less than half of the group had seen any improvement in the ulcer and one in seven people required part or all of their foot amputated.[16]

While Peter's 'Charcot' left foot is fairly uncommon, there's nothing unusual about the painless ulceration and eventual loss of his right foot. There are probably half a billion people with diabetes worldwide,[17] and roughly half of these will have damage to their peripheral nerves.[18] In some, this nerve damage causes constant pain. Others, like Peter, have a reduced or completely non-existent ability to detect tissue damage. This unwitting neglect causes and worsens tissue damage and foot ulceration, often leading to amputation (diabetic foot ulcers being the leading cause of non-traumatic amputations)[19] and also hastening death. The costs are societal as well as individual: NHS England's expenditure on diabetic foot care in 2015 was more than the cost of treatment for breast, prostate and lung cancers combined.[20] Naveed's and Jo's painlessness reveals how danger-detection works and why it is so important for surviving and thriving. In Peter's case, the devastating effect that acquired analgesia in a small, neglected part of the body can have was writ large. He is one of millions, part of a painless pandemic.

Anna

Paul Schilder and Erwin Stengel had seen everything. For psychiatrists with a penchant for research, Vienna in the early twentieth century must have been paradise. Home to Sigmund Freud, the city was not only the cradle of psychoanalysis but also the scene of numerous breakthroughs in neurology and other fields of medicine. World-renowned doctors flocked to Vienna, as did unusual patients.

Schilder and Stengel were used to having the weird and wonderful brought before them, but when they were called to assess Anna, she had them baffled. The woman was known to repeatedly injure herself if left alone. She would carelessly stab herself with knitting needles and, seemingly out of curiosity, push objects into her eyes. At first glance it seemed that Anna was showing self-harming behaviours, but when Schilder and Stengel began to examine her systematically, it slowly became apparent that the problem was with her pain:

> Pricked on the right palm, the patient smiles joyfully, winces a little, and then says, "Oh, pain, that hurts."[21]

While Anna seemed to feel pain and was aware of experiencing it, and even though her face often grimaced and winced with it, it didn't distress her and she never found it unpleasant. She wasn't bothered by the pain at all. She could describe its intensity and quality – distinguishing between a prick on the arm or a powerful electric shock – but had no emotional response to it. She showed no fear, aversion or disgust when in pain. Without an emotional reaction to pain, her withdrawal from dangerous stimuli was either slow or absent, often leading to injury. Anna was not inattentive or cognitively impaired. She was also not masochistic, deriving no pleasure from painful stimuli, either. Pain simply had no meaning for her.

Over the coming decades a few similar patients were identified, most notably by the Spanish neurologist Marcelo Berthier, who described another six cases in the 1980s.[22] Imaging and autopsy revealed that most of these patients had damage to areas of the brain that can loosely be described as the emotional brain – the amygdala, insula and anterior cingulate cortex. As the damage was most commonly caused by either strokes or brain tumours in adulthood, most patients had spent the majority of their life feeling pain 'normally'. This bizarre and exceedingly rare condition – of which Anna was the first recorded case – is called pain asymbolia. The unusual symptoms caused by damage to these specific areas of anatomy provide a vivid demonstration that pain is emotional as well as sensory. Pain asymbolics have a normal 'sensory-discriminative' pathway, which quickly conveys information up the spinal cord (in a bundle of fibres called the spinothalamic tract) into areas of the brain that identify where the damage is coming from and what kind of damage is occurring. A second, slower wave of danger transmission then travels into the brain and activates its emotion-regulating areas; damage to specific areas of these emotional structures causes pain asymbolia. This has recently been confirmed by a converse scenario. A fascinating French study, published in 2020, identified four individuals with epilepsy whose seizure activity was limited to small areas of the emotional brain. During seizures, they would feel the agonizing unpleasantness of pain and exhibit pain behaviour, but they were not able to describe what the pain felt like, nor where it was coming from.[23]

While pain asymbolics have an intact sensory pain system but a broken emotional system, there are also a tiny number of individuals who have the opposite problem.[24] If they burn their hand, they have an unpleasant emotional reaction to the pain but are unable to describe the quality of the pain (i.e. whether it is burning, stinging, sharp or crushing), nor locate its origin on the body. In these cases, the brain damage is located in areas that form the sensory pain pathway, such as the primary and secondary somatosensory cortices.

Anna was not, like Naveed, Jo, Candice or Peter, *insensitive* to pain, but *indifferent* to it. Pain asymbolics are extraordinary people who show that the transmission and detection of normal danger signals do not truly produce the experience of pain. Pain is often – incorrectly – viewed and described as a sensory experience reflecting tissue damage. But Anna showed that our perception of pain isn't merely sensing something; it is *feeling* it. Asymbolics take us to the heart of the nature of pain: it is not just a sensation, nor just an emotion, but a remarkable blend of senses, emotions and thoughts. It truly is a compelling experience – one that urges us to protect our body. Just as Beethoven requires dozens of musicians to play an assortment of instruments simultaneously to produce the transcendent experience of his finest orchestral music, we need multiple nerves and brain regions to instantly coordinate various aspects of our being to produce the phenomenon of pain. A stubbed toe, then, is worthy of a symphony.

Painless people, like this extraordinary 'painless five', are the subject of intense pharmaceutical interest. They are rapidly expanding our understanding of pain, and reveal tantalizing mechanisms for future pain-relieving drugs. They also do something simple yet astounding: they demonstrate pain's key role as a protector. Without pain, we would live lives of unprotected injury and early death. Pain wants to look after us. Even in our search for ways to soothe the life-wrecking effects of persistent pain, this truth is of the greatest importance.

3

Do I Have Your Attention?

The power of distraction and imagination

One good thing about music, when it hits you, you feel no pain.

BOB MARLEY

I LAY FLAT, MY head pressed down against the roof of the train. On either side of me, greens and browns of the wooded German countryside flew past. Ahead, a steady cloud of soot billowed out of the main engine, many carriages away. Keeping a low profile, I slowly commando-crawled forward, my eyes scanning the roof of the next carriage for my target. And then I saw him. Partly obscured by a protruding air vent was the grey coat and peaked cap of the Nazi SS officer. But he had already seen me. 'Sound ze alarm!' he screamed before raising his Luger and firing at me.

I'll come clean. At this point I was a few hours into *The Great Escape* video game on one nameless day during the long school summer holidays. But I could have been that escaped POW desperately trying to cross the Swiss border. The immersion was deep and I had no concept of time or what the rest of my family was up to. As I – or my character, as we were now one and the same – slowly inched my way up the carriage shooting Nazis and ducking under bridges, I had a vague sensation of something odd and uncomfortable happening to my left foot. But it never warranted enough attention to drag my focus away from getting to the front of the train. When, an unknown period of time later that day, I completed the level, I slumped back in my chair and closed my eyes. And that's when the pain started. I shook my left foot, from where the discomfort was emanating. Not only did my flip-flop fly off, but so did one of our newly bought kittens, which had clearly been attached to my big toe. Looking down, I saw small streams of fresh blood dripping from the toe onto the carpet. The kitten had clearly been gnawing and scratching my foot for at least half an hour, and I hadn't noticed it at all. My immersion in another dimension had completely stopped the emergence of pain in my brain, albeit for as long as the experience lasted.

The element of distraction has, appropriately, not gone unnoticed by pain researchers. Over the past twenty years, the Human Interface Technology Lab (HITLab) at the University of Washington has been using the power of virtual worlds to reduce pain. It was in this lab that virtual reality (VR) pioneer Hunter Hoffman created a place to which burns sufferers could escape while undergoing their notoriously painful wound care: SnowWorld. The patient dons VR goggles and noise-cancelling headphones, and is immediately transported from the surgeon's table to another dimension. In this VR video game, the user floats above an ice canyon in a moonlit, wintry world. Woolly mammoths wallow in the deep snowdrifts, penguins stare, confounded, down at them from the steep walls of the canyon and hordes of animated snowmen line the valley floor. The aim of the game is to hurl snowballs at the surrounding fauna, which explode in a satisfying flurry of snow when hit. All the while, the jaunty, upbeat melody of Paul Simon's 'You Can Call Me Al' plays through the headphones. This choice of song is too wonderful to be a footnote: Paul Simon once tried SnowWorld at an exhibition, and after liking everything but the music Hoffman was using, offered the use of his own. Hoffman's lab has found that when soldiers with burns injuries are transported to SnowWorld, they report a 35–50 per cent reduction in pain during procedures.[1]

These reported experiences are complemented by fMRI scans showing reductions in pain-related brain activity. Hoffman and his team argue that the remarkable reduction in pain is due not only to the element of distraction, but also to the patient's willingness to suspend disbelief: the brain is motivated to enter the game and become distracted, because the reward is reduced pain. Early evidence also suggests that combining VR with hallucinogenic drugs could further improve the pain relief by deepening the patient's immersion.[2] Compared with medications, whose pain-relieving powers have an imprecise duration (not to mention frequent side effects), VR can be targeted to a specific time-frame, which means its effects can almost immediately be turned on or off. VR works, and what is also exciting is that there are market incentives to drive the technology forward. Unless new pain treatments involve novel drugs, there is little incentive outside universities for them to be developed and distributed. In the case of persistent pain, the most effective treatments we have are often non-pharmacological. But, while VR is clearly effective in acute pain, there's just not enough evidence at the moment for any sustained effects in persistent pain.[3] But I like to speculate: imagine a future in which persistent pain-sufferers can afford a VR headset to have at home and can undergo regular immersion and distraction in a world specifically tailored to their form of pain; distraction that helps rewire their brain to feel safer and cope better with pain even outside the simulated reality.

In the real world, almost all of our pain experiences are governed by attention and distraction. When you're immersed in an exciting movie or stuck in a good novel, you don't pay much attention to everything going on around you. As you read this book, your brain is filtering out background noises, objects in your peripheral vision and even the feeling of the book in your hand. If you now fully concentrate on the weightiness of the book and the texture of its cover, you begin to feel it. All of this tactile information was already being detected by your brain as you held the book, long before you *felt* it. The extreme end of sensory distraction can be found in what has been called 'the most British conversation ever'. Lord Uxbridge, a British aristocrat and officer, was riding alongside the Duke of Wellington at the Battle of Waterloo in 1815. Lord Uxbridge had led a number of cavalry charges at the French, with the enemy's canons whizzing overhead, blasting into his soldiers either side of him and bringing down a series of eight horses from underneath him. Exhausted but utterly focused on the task in hand, it probably took him a while to notice that one of the French cannonballs had completely shattered his right leg. His words to the Duke should be read in as smart an English accent as possible: 'By God, sir, I've lost my leg'; to which Wellington replied, 'By God, sir, so you have!'[4]

This story may well be apocryphal, but from many conversations I've had with soldiers in one of the world's leading military hospitals, in Birmingham, even the most devastating injuries can feel painless in the heat of battle. This incredible phenomenon – one of the clearest demonstrations that pain is not a direct measure of injury – has been known since humans started fighting each other. The Ancient Roman philosopher Lucretius recorded how, when 'the scythed chariots, reeking with indiscriminate slaughter, suddenly chop off the limbs' of men, the 'eagerness of the man's mind' means that 'he cannot feel the pain' and 'plunges afresh into the fray and the slaughter'.[5] Lord Uxbridge felt no pain in his shattered leg partly due to distraction and an intense focus on fighting and surviving.

It is important at this juncture to debunk the myth that in these 'fight or flight' situations it is adrenaline that inhibits pain, by blocking danger signals. Adrenaline can have the effect of intensely focusing the mind so that it is distracted from pain, but it doesn't block pain and it can even increase it. The main molecules that lessen pain are opioids: our body's equivalents of morphine, stored away in our brain's drug cabinet for when the brain decides to reduce pain. Lord Uxbridge's complete lack of pain in the heat of battle was caused by something slightly more complex than pure distraction: his unconscious brain – his

Ministry of Defence – had assessed the visual and nociceptive inputs and made the decision that fighting for his life was more important for his immediate survival than feeling pain, so any danger signals travelling up to the brain from his leg were completely blocked.

Conversely, if our attention is more focused on a potentially harmful stimulus, we feel more pain. In 1954, the British psychologists Hall and Stride found that for anxious subjects the simple inclusion of the word 'pain' in a set of instructions made an electric shock agonizing, when it previously had not felt painful at all.[6] One often-overlooked takeaway from attention is that its diversion (i.e. distraction) is a potent pain reliever. Any parent can tell you this, but it can be readily used to help with all manner of pain. So let's use it. While new advancements in VR technology show lots of promise, there are many low-tech ways in which healthcare professionals and caregivers can alleviate pain. A study using fMRI to monitor subjects' brains found that, if they were given a memory task while being exposed to a painful stimulus, they perceived less pain.[7] The fMRI images showed that this distraction blocked incoming danger signals from the spinal cord. Interestingly, when the patient was given the memory task but also given the opioid blocker naloxone, the analgesic effects of distraction were reduced by 40 per cent. This shows that when we are distracted from pain, our brain's drug cabinet opens and doles out a hefty dose of opioids, blocking the stream of nociceptive signals travelling up the spinal cord. When I go to take blood from a patient, the simple act of asking questions that require thought and engagement – whether it's about their past medical history, their favourite holiday destination, or whether they would rather own a flying carpet or an underwater car – means I know that I am giving real pain relief. Similarly, engaging in activity that distracts from pain – whether it be music, reading or meeting up with others for a chat and a cup of tea – needs to be encouraged and made more readily available for those living with persistent pain.

'Close your eyes and breathe normally . . .'

Here was I – a doctor trained in the Western school of rational enquiry, empirical evidence-gathering and, dare I say it, snobbish cynicism – getting hypnotized. I'd lived with irritable bowel syndrome (IBS) for as long as I can remember, with abdominal cramps and contractions bringing regular discomfort and occasional agony. Medications and dietary changes had done absolutely nothing. IBS is a complex, poorly understood condition, but I was reading that a lot of the evidence indicated that it was caused by a dysfunction of the gut–brain axis. Paper after paper I studied lent more weight to the argument that an irritable mind (in my case, from mental stress from academic studies and hospital shift work) can lead to an irritable bowel. My research kept bringing up what appeared to be an effective treatment for IBS – one that, despite six years at medical school, I had never come across: hypnotherapy.

Any prior knowledge of hypnosis I had was either comic or sinister – from pocket-watch-swinging stage hypnotists to mind-controlling movie villains. Thankfully, Paul, my hypnotherapist, was neither of these. A genial man in a lively shirt, Paul led me to his clinic room on the outskirts of Oxford, a calming arrangement of muted pastel greys and soft furnishings. Warm, early autumn light filtered through the blinds. There was not a pocket watch to be seen. I took my seat in a comfortable, pewter-coloured armchair opposite Paul's sofa and rested my feet on the thick faux-fur rug. Paul had fallen into hypnotherapy as a consequence of it transforming his own life, he told me. He'd had a traumatic upbringing; raised by a mother with bipolar disorder and alcoholism so uncontrolled he constantly feared her imminent death. He found counselling transformative, but experienced something deeper and even more powerful when he underwent hypnotherapy to try to kick his smoking habit. Paul vowed to learn how to wield this powerful tool himself, spending years studying it and slowly building a hypnotherapy practice treating everything from anxiety to addictions, from phobias to pain.

'Become aware of the way your chest and stomach move as you breathe in and out . . . Begin to notice that, as you breathe, your shoulders rise up, and as they rise up, they pull upon your biceps and triceps. Notice them all at once, relaxing as you breathe . . .'

Through words of suggestion, Paul gently guided my attention across my body; I felt as though I was selectively shining a torch on sensations I never pay any attention to: the heaviness of my feet, the sound of my breathing, the difference in temperature between the back of my neck and the back of my throat.

'Now begin to hear all the other sounds going on around you . . . Ever since you came into this room, you've heard that clock ticking faintly on the wall, you've heard the whir of the air conditioning, traffic outside. You hear everything going on around you all day, but you haven't been paying attention to these sounds. Your brain wants you to focus on things it believes are more important, things you've trained it to think are more important.'

I was now completely relaxed and yet hooked on his every word. Paul directed my focus to my thoughts, whether good, bad, stressful or relaxing. He told me that I didn't need to pay undue attention to any particular thought; I could notice, observe and accept it, or I could treat it as noise, just like the ticking clock on the wall. Then he took me back to my bodily sensations, drawing my attention down to my sore, cramping abdomen, and used imagery to change the way I saw my pain.

'Picture your intestines . . . Look at them as though they are a river . . . Now, it might be a fast stream of rocky rapids, but imagine it as the gentle Thames as it flows past Oxford, with languidly moving punts gently drifting downstream.'

Hypnosis is one of the oldest talking therapies (known in the medical world as 'psychotherapies') in the West. But it got off to a really bad start and it hasn't fully recovered. The origins of modern hypnosis are usually attributed to Franz Mesmer (from whom we get the verb 'to mesmerize'), an unconventional eighteenth-century German doctor. Mesmer believed that all living things are connected by an invisible natural force he called 'animal magnetism', and that all illnesses stem from an interruption in its flow. He would sit opposite a patient, his knees touching theirs, and would melodramatically pass his hands over their body, sometimes for hours, until the patient experienced a crisis – usually collapsing or convulsing – which he said brought about the cure. When Mesmer moved to Paris, he rapidly gained celebrity status as well as the ire of the medical and scientific community. In 1784, a panel of nine top scientists – which included Founding Father Benjamin Franklin and Antoine Lavoisier, who discovered oxygen – was appointed by King Louis XVI to investigate animal magnetism. In a series of ingenious experiments – and one of the earliest examples of evidence-based medicine – the dream team completely debunked Mesmer's theory. But they may have made the same mistake doctors make today: just because we can't exactly explain how hypnosis works (and some explanations for it do seem scientifically dubious), we often miss the crucial fact that, in many cases and for many people, it does work. Benjamin Franklin's summary is revealing: 'the practice of magnetisation is the art of increasing the imagination by degree'.[8] Animal magnetisation is clearly bogus, but if we can ease pain and suffering by harnessing the power of imagination, then why not?

It is only in recent years that hypnosis has been spilling its secrets and is starting to be recognized as a scientifically legitimate treatment for a number of conditions, including some forms of pain. But hypnosis' image isn't helped by the fact that it is still very hard to explain. If you asked ten hypnotherapists and ten scientists researching hypnosis to define it, you'd probably get twenty different definitions. Most definitions riff on the theme that hypnosis is an altered state of consciousness that makes someone more responsive to suggestion. Under hypnosis, people have their attention intensely focused on one area or idea, and this absorption in a subject can reduce or shut out their peripheral awareness; in a sense it is a form of distraction. The controlled focusing of attention minimizes competing thoughts and sensations. One word that is often used in hypnosis is 'dissociation', which is where different parts of our awareness are separated from each other: our unconscious brain can listen to and respond to suggestion without the awareness of our conscious mind. It also explains how some people under hypnosis can obey commands without being consciously aware of it.

You probably dip into a state of hypnosis more than you realize. Think about how many times you have driven home from work and not remembered anything about the commute – it's as though another, unconscious driver inside you was at the wheel. Or when you're at a noisy party but your ears seem to selectively tune into the voice of the person you're chatting to, filtering out all of the other sound – until a person in a nearby conversation mentions your name and the spotlight of your attention suddenly falls on them. It's as though someone in your brain, hidden from your conscious awareness, was listening to that conversation the whole time.

Most hypnotherapists agree that hypnosis requires two key stages. During the 'induction' phase, Paul told me to close my eyes, relax, and completely focus my attention on what he was saying. In the 'suggestion' phase, he used words and imaginative imagery to help me look at, and deal with, my pain from a new perspective. Suggestion helps the subject to experience imaginary events and ideas as though they are real. One of Paul's favourite definitions of hypnosis is 'believed-in imaginings' and, to show me the power of imagination to influence mind and body, he made me take what is known in the hypnotherapy world as the 'lemon test'. As long as you can safely close your eyes, you can take this test too. Read the following passage carefully, then close your eyes and imagine the same scenario in vivid detail:

> Imagine that you are standing in your kitchen at home. Take in the usual sights, sounds and smells. Then walk over to your fridge, open the door and, on the bottom shelf, imagine a lemon. Imagine this to be the perfect lemon – the perfect shape, the perfect colour of yellow. Grab hold of the lemon and feel its weight and waxiness in your hand. Bring it up to your nose and sniff the fresh scent of the lemon. Place it on a chopping board, pick up a knife and slice it in half. Pick up one half of the lemon and imagine a small amount of juice running over your fingers as you bring it up to your mouth. Now bite into it, and taste all of the sour, tangy juices on your tongue.

If your mouth has begun to water, it indicates that you're towards the more hypnotically suggestible end of the spectrum. Importantly, for the study of hypnosis and its use in treating disease (hypnotherapy), we all respond to it differently. Roughly 10–20 per cent of people are highly hypnotizable, a similar percentage doesn't seem to respond at all, and everyone else falls somewhere in between. But, for those who respond at least partially, suggestion and imagination taps into our subconscious in a way that can change both mind and body. If, instead of imagining the lemon, you simply tried to consciously instruct your salivary glands to activate and secrete saliva, they wouldn't have obeyed.

I left my therapy session with Paul feeling calm and confident. I set aside ten minutes each morning to sit, close my eyes and listen to the audio recording of our session. Over the following couple of weeks, I could still feel painful sensations during flare-ups, but the pleasant imagery I was now associating with IBS was beginning to change my experience of it; the lens through which I was experiencing the pain was making it seem less and less negative. It was as though I could take a step back and look at the pain as an observer, rather than viewing it as something threatening. A large proportion of Paul's clientele are looking to overcome phobias; I wonder whether a similar process was at work in me. Pain is like an ugly, threatening-looking spider, which I slowly learned to view as a harmless friend: instead of fleeing into another room or trying to thwack it with a newspaper, I could now gently pick it up and rehome it in the garden. Weeks, and then months, went by and, as I write, my IBS has completely stopped rearing its head.

Hypnotherapy seemed to work for me, and it clearly works for others too: robust studies show it gives significant pain relief in 50–75 per cent of IBS sufferers.[9,10] After my experience, I felt that there was something special, something unusual, about hypnosis, something that could transform our understanding of pain and relieve many people living with persistent pain. But I still had plenty of medical colleagues who argued that hypnosis is not special at all, that it is not a distinct psychological state and that hypnotherapy's effects can be explained by other means. A common assumption is that hypnotherapy works via the placebo effect. A persuasive therapist with soothing words in a dimly lit room enhances the expectation for something to work – essentially a self-fulfilling prophecy. Another theory is that being induced into hypnosis is just the process of adopting a role and responding to social pressure and cues. Or perhaps hypnotherapy eases pain in some people through relaxation. But a large body of evidence – ranging from whacky but ingenious experiments to state-of-the-art neuroimaging – suggests that there is something deeper going on in the brain.

Some of the first studies on hypnosis in the context of pain found that there certainly was something special about it. In a neatly designed 1969 study by psychologists at the University of Pennsylvania, participants were subjected to muscular pain through the inflation of a blood-pressure cuff. The researchers found that – particularly in those who were easily susceptible to hypnotic suggestion – pain perception during hypnosis was reduced and pain thresholds were significantly increased.[11] This also far exceeded the pain relief participants felt when they were later given a placebo that they had been told was a powerful pain-relieving drug. The placebo effect certainly accounts for an element of the effectiveness of hypnotherapy, but there is something else going on as well. A study a few years later revealed a startling – albeit somewhat spooky – potential explanation for this. Highly hypnotizable people are able to respond to questions by 'automatic writing', where one hand writes answers to questions without the subject's awareness. In 1973, the renowned Stanford psychologist Ernest Hilgard tested this out on a young woman – let's call her Lisa – by first asking her to rest her hand in ice-cold water.[12] Unsurprisingly, she found this intensely painful. Lisa was then induced into a hypnotic state, and again her hand was placed into the ice water. This time she reported feeling no pain whatsoever, but while she was verbally describing how relaxed she felt, her other hand continued to automatically write, reporting that she was feeling agonizing pain – the same pain she felt when she was not hypnotized. Her body also showed other signs that she was in pain and distress – a raised heart rate and increased blood pressure – though she did not consciously feel any pain. Hilgard inferred that the pain divided – dissociated – her consciousness, so that some elements of her brain were reacting to the danger signals travelling from her arm, but the elements of her brain producing her subjective awareness felt no pain at all.

At the turn of the twenty-first century, a group of scientists from Harvard and Stanford made use of new technology to unveil the power of hypnotic suggestion.[13] Participants in one study were shown a series of images, either in colour or in greyscale. Eerily, when participants were put under hypnosis and were told that the greyscale images were coloured, they saw colour. When they were shown brightly coloured images but the verbal suggestion was that they were seeing dull greyscale, they again saw what the hypnotists told them. Importantly, though, this response was only seen in highly hypnotizable participants. What gave this study extra experimental oomph was that during hypnosis the participants' brains were scanned using fMRI machines. The scans showed that the areas of the brain that process colour images were activated whenever the hypnotized subjects were told they were seeing colour. Under hypnosis, believing is seeing.

In the couple of decades since that study, brain imaging has helped us further peer into the mysteries of hypnosis. In a 2016 investigation, a team led by Dr David Spiegel, Professor of Psychiatry and Behavioral Sciences at Stanford University, explored the brains of highly hypnotizable people in fMRI scanners.[14] They found three key hallmarks of the hypnotic brain. First, hypnosis decreases activity in the brain's salience network: you are so absorbed and focused on one thing that you ignore everything else. This also happens when you're engrossed in an exciting movie, book or video game and don't pay much attention to your surroundings. Second, Spiegel's team found a reduction in connections between two brain areas (the prefrontal cortex and the default mode network), representing a decoupling of a person's actions and their awareness of those actions; in a hypnotic state you don't think, but just do – just like your automatic commute home, driven by an 'unconscious' driver. Finally, and most surprisingly, the team saw an increase in connectivity between two specific brain areas (the prefrontal cortex and insula) that suggests an enhancement of mind–body control. Rather than hypnosis being about losing control – driven by the classic image of a power-hungry, mind-controlling hypnotist trying to make people collapse to the floor or cluck like chickens – hypnosis actually improves it. It enables someone to have more influence over their perceptions – including pain. This third finding may well be behind the effectiveness of hypnotherapy in individuals with persistent pain, and shows that it is not just a case of deep distraction.

And fMRI imaging has confirmed this effect of hypnosis, of breaking pain circuits in the brain. In 2005, researchers at the

University of Iowa found that patients exposed to painful heat under hypnosis experienced greatly reduced pain and showed reduced activity in brain regions associated with pain networks, most notably the primary somatosensory cortex – the area involved in localizing pain.[15] Interestingly, there is evidence that hypnosis can reduce the perception of pain by quietening either the sensory or emotional areas of the brain, depending on what words the hypnotherapist uses. One study, at the University of Montreal, found that hypnotic suggestions directed at the emotional experience of pain resulted in participants experiencing a heat stimulus as less unpleasant, though there was no change in pain intensity.[16] When the suggestions were focused on the sensory elements of the pain, the opposite happened: the pain became less intense but retained its unpleasantness. It seems that effective hypnotic suggestions should aim to tackle these two elements of pain.

We're still a long way off from a complete understanding of the neuroscience behind hypnosis. Debates will roll on, and the practice may never shrug off its sense of the mystical. But there's one thing we do know: for many people, in many types of pain, hypnotherapy works. A broad evidence-base shows that hypnosis and hypnotherapy are scientifically proven ways of reducing short-term pain, and for some it really can reduce long-term pain.[17,18] Hypnosis is effective at reducing procedure-related pain in children[19] and studies show that it reduces the need for opioid-drug use and even the need for general anaesthetic in surgical patients, at minimal cost.[20,21] 'Hypnobirthing', where hypnosis is used to relieve the pain of childbirth, is also supported by solid evidence.[22] A long-term study of children with irritable bowel syndrome and functional abdominal pain (pain that cannot be explained by any measurable means) found that hypnotherapy not only improves pain but that its effects are long-lasting.[23]

These studies tell us that hypnotherapy is not just a placebo, but can change patterns of thinking, training the brain out of pain. The use of hypnotherapy for IBS, even when all conventional medical treatments have failed, has been so successful that the UK's National Institute for Health and Care Excellence (NICE) has given hypnotherapy its seal of approval – a very rare boon for a 'complementary' therapy. A randomized controlled trial published in October 2020 found that hypnotic cognitive therapy (hypnosis combined with cognitive behavioural therapy) was very effective at reducing persistent pain, and more effective than either hypnosis or cognitive behavioural therapy on their own.[24] Hypnotherapy can also relieve pain indirectly by targeting conditions that worsen pain, such as anxiety, insomnia and post-traumatic stress disorder (PTSD).

After a few sessions with a hypnotherapist, it is easy to learn self-hypnotherapy, which (if practised regularly) can change one's relationship with pain. At its most effective, self-hypnosis can completely eliminate the experience of acute pain. There are groups of seemingly superhuman individuals across the world who can manipulate their awareness to such an extent that they shut off danger signals travelling up from the peripheries, from Greek fire-walkers to Nordic ice-swimmers. A 1982 study examined an Indian fakir who, after two hours of intense meditation, could enter a phase of thinking about nothing, during which he could pierce his body with pins and daggers while feeling absolutely no pain.[25] His electroencephalogram (EEG) brain scan in this trance state showed significant 'theta wave activity', brain oscillations associated with deep concentration and memory retrieval, also something consistently seen across people undergoing hypnotherapy.[26] This fakir's superpowers are beyond the reach of most of us and require years of training, but they show the brain's powerful influence over pain.

The caveat is that there is significant variability in how people respond to hypnotherapy, and it certainly doesn't work for everyone. We clearly need these therapies to be stratified and personalized – working out who is likely to respond to hypnosis before recommending it as a therapy, and offering other treatments for people who would likely be non-responders. Nonetheless, while hypnotherapy is sadly undervalued, under-investigated and underused in mainstream medicine, I've become convinced that this mind–body therapy enables a great many people living with long-term pain to slowly disassemble the mental structures of persistent pain, and begin to live fully again.

Perhaps hypnotherapy's greatest problem is its image. On a surgical ward-round during one of my first placements as a junior doctor, we discharged a young man who was initially suspected to have appendicitis; however, when it became apparent he was instead suffering a very bad bout of IBS, the surgeons had completely lost interest. The patient had previously undergone many medical investigations and treatments, none of which had worked. When I suggested to the consultant that we could get him on the waiting list for hypnotherapy (I had recently looked up the evidence while investigating my own IBS), he smirked and made sure his trainee surgeons were in earshot: 'Aha! I see that Freud's got the diagnosis! Why couldn't we see that his abdominal cramps are latent sexual longings for his mother?' His eyes darted around the sleep-deprived gaggle of surgeons to ensure that he'd extracted sufficient polite laughter out of them. Although most doctors don't see hypnotherapy as outright quackery, they continue to argue that there aren't enough randomized controlled trials to form an evidence-base. When it comes to new drugs, such trials are the gold standard: neither the doctor nor the patient knows which tablet is the drug and which is the placebo, eliminating the power of the placebo effect. But this can't be applied to treatments such as hypnotherapy, given that you can't hypnotize someone without their knowledge, and that suggestion and expectation are central to the success of hypnotherapy. It's also unsurprising to find that, as hypnotherapy cannot be patented, there isn't a huge drive to fund studies into its effectiveness as pain relief.

As hypnotherapy and virtual reality continue to show their worth in the lab and in the clinic, it's exciting to imagine what might happen when they are combined. Early results in some pioneering centres look promising. Anaesthetists and hypnotherapists from Strasbourg, France, teamed up to create a system they call HypnoVR. Patients don VR goggles, choose where they want to be transported to (either a quiet forest, an underwater world, a tropical beach or the stillness of space) and are then induced into hypnosis by the soothing tones of a voice reading a script created for the patient's specific medical procedure or pain. This can be used during procedures or for pain experienced immediately after surgery. Children given a twenty-minute HypnoVR session within seventy-two hours of surgery required half the amount of post-operative morphine compared with those who received standard care, and their hospital-bed time was reduced by twenty-one hours.[27] Hypnosis is one of the world's oldest psychotherapies and virtual reality is one of the world's newest technologies. Both clearly show that attention, distraction and imagination are critical pieces in the puzzle of pain. Their manipulation provides real hope for pain relief. In a world where the only analgesic options presented are those that numb sensations with pills or procedures, these therapies encourage a playful, powerful re-engagement with the world.

4

The Expectation Effect

Placebo, perception and prediction

He cures most successfully in whom the people have the most confidence.

GALEN, SECOND CENTURY AD

PAUL EVANS WAS thrilled to finally have his hands on an electrical stimulation machine. The radio producer had been suffering with agonizing and exhausting fibromyalgia for years. Flare-ups are common and debilitating: 'I just feel awful. I'm aching in every joint in my body.'[1] He wanted something to relieve the pain and had seen evidence for the effectiveness of electrical stimulation devices, as well as hearing good things about them. During one of his interviews on *Airing Pain*, the radio programme of the British charity Pain Concern, he described the joy he found in attaching the pads to his skin. 'When I bought it, it was absolutely fantastic! It was like taking marijuana . . . I've never taken marijuana, but if this is what it's like then, yes . . . It just relaxes me.' It appeared that Paul had found an electrical elixir of pain relief. But then he went on: 'After three months I found that I hadn't plugged it in.'

Paul had unwittingly discovered the power of placebos. The placebo effect, like pain, is strange. It is a window into the true nature of pain. It shows that pain is the product of a context-based decision made by our brain – usually without our knowledge. To truly understand the wondrous weirdness of the placebo effect (and its relevance for pain in general), we first need to define our terms. A placebo is essentially something that appears to be a medical treatment but is in fact inert or inactive, such as a sugar pill, a saline injection or Paul's switched-off stimulation machine. The placebo effect is the response our brain makes to the *context* in which the treatment is delivered.

The word placebo derives directly from the Latin (meaning 'I shall please') and has something of a bad rap in the English language. In medieval Europe, 'placebo' was a word used in songs chanted at funerals and it soon became associated with funeral-crashing con artists who would 'sing placebo' to get a share of the funeral meal. Any use of the word soon became synonymous with sycophants. *The Merchant's Tale* – a fourteenth-century story from Geoffrey Chaucer's *Canterbury Tales* – includes a character called Placebo: a fake, flattering yes-man who lets his brother make terrible decisions without giving him any advice or resistance.[2] Although the word is no longer used to describe flatterers, its connotations of fakery have transferred to the medical world.

The almost supernatural ability of inactive medications or sham treatments to affect our body and relieve pain has confounded scientists for centuries and shrouded the placebo effect in mystery and suspicion. A paper published in *The Lancet* in 1954 asserted that placebos are simply a mental crutch and comfort, particularly to 'unintelligent, neurotic, or inadequate patients'.[3] Today, some doctors argue that the placebo effect itself is fake, with patients simply being fooled into thinking that they are better, when in reality they are not.[4] Most modern evidence, however, argues against this: placebos can change our brain and genuinely alter symptoms and diseases modulated by the brain. A slew of fascinating studies carried out over the past couple of decades is showing that the placebo effect gets to the very heart of the experience of pain, and could revolutionize our treatment of it.

A placebo doesn't have to be just a sugar pill or a saline shot.

Around the turn of the millennium, a daring and inventive group of Houston-based orthopaedic surgeons decided to carry out 'placebo surgery', in what has become a landmark study.[5] At the time, one of the most commonly performed orthopaedic operations was 'arthroscopic debridement', which involves making an incision in the knee, opening the knee joint and removing inflamed tissue and bits of loose cartilage and bone. It is intended to reduce the pain of knee osteoarthritis and, in many cases, is successful. But the surgeons admitted that no one really knew how or why the procedure works. They decided to split a cohort of 180 patients with painful knee arthritis into groups. One of these groups had the usual surgery. Another group underwent sham surgery: they were put under general anaesthetic, an incision was made in the skin and that was it. Remarkably, the placebo surgery was just as effective as the real operation. Even more remarkably, over the two years of follow-up (where the patients were still in the dark as to whether they'd had the real surgery or the placebo) the placebo group reported better outcomes. The study showed that the pain relief brought about by this procedure did not come from changing the tissues, but from changing the brain through expectation and hope.

Recent studies support these findings, even suggesting that exercise is as effective at treating osteoarthritic knee pain as surgery (apart from specific cases where ligament damage requires operating).[6] Placebo surgery is certainly not limited to arthroscopic debridement of the knee; a 2014 review paper found that in fifty-three trials where surgical procedures have been compared to placebo alternatives, in half of the cases the fake surgery was just as effective as the real procedure.[7] One thing we could take from these studies is an assumption that those 'real' operations were ineffective. But they weren't; it's just that they worked in the very same way as the placebo procedures: not by addressing internal tissue damage but by tapping into the placebo effect. If we could harness the power of placebo (perhaps without the risks of fake surgery), it could make a real difference in pain medicine. We need to find out how, and why, giving 'nothing' works.

In 2004 Tor Wager and his team at Columbia University wanted to use neuroimaging to find out what really goes on in the brain during the placebo effect. The team subjected volunteers to electric shocks while observing their brain activity in an fMRI scanner. Unsurprisingly, pain-sensitive areas of the brain (such as the thalamus, anterior cingulate cortex and insula) lit up when the shocks were given. When the participants were given a placebo cream on the skin before the shocks were applied, which they were told would reduce the pain, the activity in these pain-related brain areas reduced.[8] The correlation was there; so far, so good. But Wager wanted to know *how* this pain relief occurred. A few years later, his team used positron emission tomography (PET) scans to measure the activity of opioid receptors in the brain.[9] They found that a placebo treatment increased the release of opioids across a range of pain-relevant brain areas. The mere expectation of pain relief – the belief that you will feel less pain – is enough for our brain to unlock its own drug cabinet and dispense a strong dose of wonderful opioids such as endorphins – essentially, non-addictive morphine. Wager's experiment helped to confirm previous studies, including a seminal study in 1978 that demonstrated that you can reverse the placebo effect by giving someone the opioid blocker naloxone.[10] Another study in 2009 conclusively showed that naloxone blocks opioid action in key areas of the brain that influence pain, including the anterior cingulate cortex and periaqueductal grey.[11]

Other molecules in the brain are also influenced by the placebo effect. These include the release of other natural painkillers, such as cannabinoids, our body's version of the pain-relieving ingredients in the cannabis plant.[12] When someone anticipates pain, receiving a placebo also releases dopamine – a molecule associated with motivation and pleasure – in an area of the brain called the nucleus accumbens, a crucial piece of our brain's reward circuitry.[13] The more activated this reward circuitry becomes, and the more we expect the reward of pain relief, the stronger the placebo effect. Importantly, these studies show that the placebo effect reduces pain not by tricking gullible people into imagining that they are feeling better; it genuinely releases a powerful cocktail of painkillers in the brain, and it affects the exact-same pathways exploited by active medications. When someone is exposed to a pain-causing stimulus but takes a placebo, danger signals travelling from the body to the brain are suppressed by a descending, inhibitory pathway from the brain. This natural pain-relieving cocktail is the same one that soothes a soldier's pain when they are injured in the heat of battle. There's also evidence that the placebo effect stops danger signals at the level of the spinal cord, so the signals never reach the brain.[14]

We have to remember not to confuse causality. This isn't the placebo – an inert substance – doing the work; let's give our brains the credit. It is our belief in the treatment that opens up the brain's drug cabinet. The active ingredient is *expectation*. This is neatly seen in the hierarchy of fakery; not all placebos are created equal. Saline injections tend to have a greater pain-relieving effect than sugar pills,[15] and it wouldn't be surprising if fake surgery is significantly better than both of these. An expensive placebo is more effective than a cheap one.[16] The more dramatic the intervention, the more meaning the patient attributes to the treatment; and the better the rapport between patient and treatment-giver, the higher the expectation of pain relief and thus greater the actual pain relief. This is probably why intimate, intensive therapies infused with meaning and ritual are often so effective.

A large, rigorous trial in Germany found that sham acupuncture (where the needles hardly penetrate the skin and are intentionally put in all the wrong places) is as good at relieving pain as real acupuncture.[17] But, interestingly, the study found that both placebo and acupuncture tended to be better at relieving pain than conventional treatment, which included both painkillers and non-pharmacological therapies. Even if the only active ingredient of acupuncture is belief, it clearly has pain-relieving effects for some people, perhaps particularly in those for whom conventional medicine has provided little relief. A crucially important caveat is that this doesn't mean I would recommend acupuncture as a treatment for persistent pain. It's deceptive to endorse a treatment that works only through the power of expectation, particularly a treatment that can be expensive and encourage someone in pain to be a passive *patient*, not an empowered, confident *person*. Having said that, while we strive to find evidence-based treatments that really work for people in pain, it would be remiss to ignore the power of belief, expectation and confidence, something that can be found in the pain-relieving power of the interaction between the caregiver and the person in pain. Doctors, including myself, understandably want to only use new treatments if they have been shown to perform better than a placebo in trials, whether there is some 'active ingredient' or mechanism providing pain relief. In clinical trials, the word placebo is associated with failure: if a new drug doesn't perform better than a placebo in early trials, it has effectively exploded on the launch pad. But we are missing something precious if we ignore the pain-relieving power of expectation itself. If something eases someone's pain, then it eases someone's pain – and that can only be a good thing.

The power of belief when relieving pain may also be behind another intriguing phenomenon. Jeffrey Mogil and his team at McGill University in Montreal analysed an enormous amount of clinical trial data from between 1990 and 2013. Over that period, the strength of the placebo effect increased in the USA, whereas no changes were observed in Europe or Asia.[18] The cause of this isn't clear but the growing placebo effect has been making it harder for pharmaceutical companies to prove that their new drug works. The team noticed that trials in the US got longer and larger over this period, while this didn't happen in the other countries. Perhaps long, fancy, well-funded trials increase the belief of participants in the placebo group that their pill works. Perhaps it's also because the US is one of the only countries in the world with direct-to-consumer drug advertising, influencing belief and increasing expectation. While this is a headache for the pharmaceutical world, it furthers the evidence that giving people time and attention can increase the pain-relieving effect of a treatment.

What is also becoming increasingly evident is that the person giving the treatment has a huge impact on the pain relief of the person receiving it. For starters, seeing someone give you pain relief is analgesia itself. If a pain-relieving drug is given to someone through an intravenous drip, it is 50 per cent more effective when the doctor explains what's happening, compared with it being administered by a computer and without the participant's knowledge.[19] The confidence of the person giving the medication is also crucial to its effectiveness. This was beautifully illustrated in a weird and wonderful wisdom-tooth trial in which one of three possible injections was given to patients following wisdom-tooth removal. The first, fentanyl, is a strong opioid that should relieve pain. The second, naloxone, is an opioid blocker that definitely shouldn't relieve pain. The third, saline, is the placebo. One group of dentists were told that they were giving either fentanyl or a placebo to their patient. The other group were told that they were giving either naloxone or a placebo. When the pain-relieving effect of the placebo in both groups was compared, the results were astounding. When the dentist thought that there was a 50 per cent chance the patient might get the fentanyl, the placebo resulted in a 30 per cent decrease in pain, whereas in the group where the dentists thought there was a 0 per cent chance of the patient receiving fentanyl (because the patient would get only naloxone or saline), there was a 20 per cent *increase* in pain in the placebo group.[20] Confidence is contagious; a patient can pick up the subtlest nonverbal cues, and that can strongly influence their expectation of pain relief.

Understanding which individuals and groups of people respond better to the placebo effect would also be incredibly valuable, both for better understanding placebo control groups in drug trials and – brushing ethics aside for the moment – to see whether incorporating placebos into treatment could be effective in relieving pain for some people. In the (not too distant) past days of paternalistic medicine, it was generally assumed that the placebo effect worked only on the neurotic and unintelligent. In fact, evidence shows the opposite: people who are more optimistic, reward-seeking and resilient are more suggestible to the placebo effect, due to a heightened expectation of pain relief. In 2009, the University of Manchester's Human Pain Research Group found that people with high 'dispositional optimism' were likely to have a significant placebo response.[21] Being 'susceptible' to the placebo effect isn't a bad thing at all. In fact, it is good: it leads to our end goal of more pain relief.

We've all heard about the placebo effect, but it is impossible to truly understand it (and its relevance in relieving pain) without exploring its evil twin: the nocebo effect. Nocebo – from the Latin 'I will harm' – is treatment that has a negative effect due to negative expectations. Understandably, it's difficult to get ethical approval to study this phenomenon, but it's all around (and in) us. It explains why participants in the placebo group of a drug trial get side effects, but only those of the real drug that were described to them at the start of the trial.[22] A fascinating study, the results of which were published in late 2020, found that roughly 90 per cent of the side effects experienced with statins are due to the nocebo effect.[23] The nocebo effect may even underlie episodes of 'group hysteria'. There have been numerous reports of mass fainting episodes in schools in conflict zones.[24] These are often initially put down to poisoning, but extensive medical tests have shown no evidence of foul play. Perhaps the tense, suggestible environment lends itself to psychological expectations of harm, which sometimes spill over and manifest physically.

The nocebo effect is extremely common and affects the lives of millions with persistent pain. It's incredibly easy for a doctor to trigger the nocebo effect: 'You're a high-risk patient' and 'Let me know when it starts hurting' are phrases that can make pain worse and heighten anxiety. Little words can have a big impact, and negative statements stick like Velcro. We can inadvertently use the nocebo effect on others and ourselves: 'How's your rusty knee doing?', 'My back's stuffed', 'I'm falling apart at the seams'. Googling a condition before we visit the doctor may also set our expectations for pain. As humans, we are highly suggestible at the best of times, but when our body feels threatened or harmed, negative words can have an enormous effect. Verbally induced anxiety releases neurotransmitters that can open the floodgates for pain,[25] and imaging shows that they stimulate pain networks in the brain.[26] It's time to revise that wildly incorrect nursery rhyme: sticks and stones may break my bones, but words can never hurt me. Words can literally hurt us.

The good news is that positive words and suggestions also have a positive effect on pain. As a doctor, I have to (and want to) tell patients the truth (and I'm sure that pretending there won't be any pain at all could make pain worse when it comes) but that doesn't mean that I can't try to reduce a patient's pain by focusing on the positives. After assessing an injured arm, I could either leave the patient and go and write my notes, or talk to other colleagues about the case, or I could ask the patient about their good arm, reminding them that it's not all bad news. 'How's your other arm? Fine? Fantastic! Are you able to stretch it out . . . move your fingers . . . any pain? No? Great!'[27] A study at Harvard Medical School found that when women were about to have an injection of spinal anaesthetic before childbirth, those given placebo words ('We're going to give you a local anaesthetic that will numb the area; you'll be comfortable during the procedure') experienced less injection pain than those given nocebo words ('You're going to feel a big bee-sting; this is the worst part of the procedure').[28]

Maybe this seems like common sense, but I see it used so rarely in clinical practice. These positive cues can genuinely relieve short-term pain. Confident compassion is a powerful pain-relieving drug in itself. But, perhaps more excitingly, positive words, metaphors and concepts that encourage ideas of safety and lessen ideas of danger are an incredibly effective and, frankly, revolutionary way of managing long-term pain. This goes beyond placebo and involves rewiring the brain (which forms the basis of Chapter 11).

There is a simple, neat study that cuts straight through to the wonder of pain, and caused me to question my own pain beliefs.[29] Even as a junior doctor, I'd been taught that pain equals tissue damage and that you manage pain with drugs: the worse the pain, the stronger the painkiller. The study was carried out by neuroscientist and pain expert Irene Tracey's team at the Oxford Centre for Functional MRI of the Brain in 2011. Healthy study participants were hooked up to an intravenous drip and were exposed to a pain-provoking heat stimulus. Without their knowledge, an infusion of remifentanil – a powerful opioid pain reliever – had started running through the drip, and would continue to do so throughout the experiment. The remifentanil reduced their pain, but only slightly. The participants were then told that the remifentanil infusion was about to start (even though they were already on the drug). The painkilling effect doubled. The participants were then told that the infusion was about to stop, which wasn't the case. Even though the infusion was continuing at exactly the same rate, its pain-relieving effect suddenly disappeared and the participants were in pain again. That the initial, surreptitious infusion of the powerful opioid didn't produce much relief is complemented by studies showing that opioid painkillers are about a third less effective if we are unaware that we are taking them.[30] fMRI imaging showed that in the 'placebo' part of the trial, when participants were falsely told the infusion was going to start, their 'descending pain modulatory system' – areas of the brain that suppress danger signals travelling up to the brain from the body – activated. We don't actually need to take a placebo to experience the placebo effect. In the nocebo part of this experiment, where participants were falsely told the infusion was being stopped, the network by which anxiety amplifies pain activated.

Perhaps we should ditch the terms placebo or nocebo effect, and instead call it 'the expectation effect'. We can strongly manipulate the experience of pain – for better or worse – through beliefs and expectations. Our brain is immensely powerful; let's use it. One way of applying this practically is by adding meaning to the taking of a pill.[31] This could involve creating a little ritual when you take pain relievers, from simply taking them at the same time each day to spending time visualizing and imagining the improvements that the pill is going to make. Dan Moerman, Professor Emeritus of Anthropology at the University of Michigan and placebo-effect expert, even talks to his tablets: 'Hey guys, I know you're going to do a terrific job!'[32] Creating daily rituals of healthy living – from exercising to socializing to meditating – also harnesses the placebo effect in a positive way.

But perhaps the greatest implication and the biggest takeaway of modern placebo research is the importance of the caregiver, whether they be a doctor, a spouse, physiotherapist – whoever. Western medicine has made enormous progress in the realm of new medicines, techniques and technologies, but in many ways it has ignored the vital importance of the carer–patient interaction – the need to build a relationship with trust and honesty.[33] Positive information, constructive words, long and meaningful interactions really do work. This is something I think has been lost in the British primary healthcare system. Among the complex reasons for this, patients often no longer see the same family physician, have consultations limited to ten minutes and do not have the power to choose which primary or secondary care professional they will see. The very anticipation of going to see a doctor with whom you have a good rapport, or who you believe to be skilled and knowledgeable, is medicine itself. These things add up: they reduce pain by increasing feelings of safety and decreasing the perception of threat or danger; and – importantly – they provide hope.

The placebo effect seems to go against everything we've been told about the body. And yet it works. And it gets even more extraordinary, as I discovered to my delight and shock on a rainy Sunday afternoon in my late teens. Until being seemingly cured of my IBS by hypnotherapy, I lived with irritable bowel syndrome for as long as I can remember. This particular day, I had spent the afternoon lying on a sofa in the foetal position. A dear family member came to console me. They are a firm believer in homeopathy, a branch of alternative medicine that has been comprehensively debunked. Homeopathy pills are biochemically inert: the active ingredient (if there is one in the first place) has been diluted to such an extent that it no longer exists and is supposed to reside in the pill via 'water memory', an unproven concept that goes against all the laws of nature. After decades of research and countless studies, the overwhelming consensus is that homeopathic pills perform no better than placebos.[34] My family member opened a small, transparent cylindrical container – labelled with some kind of faux-Latin name I didn't bother to read – and popped a tiny sphere of what must have been sugar onto their palm. 'I know you're a bit of a sceptic and don't believe in this sort of thing, but just give it a try.'

Obliging, I picked up the ball of nothingness and swallowed it. Within a couple of minutes, the terrible cramps had completely vanished. I was ecstatic, in a world of sweet relief, and thanked them only semi-sarcastically for curing me through the placebo effect. But an unnerving thought began to seep into my consciousness, and it kept slowly gnawing away at me: the placebo effect is based on belief and expectation, so how did the sugar pill work if I knew it was a sugar pill and didn't expect it to do anything? That's when I came across the work of Ted Kaptchuck.

Kaptchuck is a professor of medicine at Harvard, but he didn't get there the usual way. After receiving his BA from Columbia University in 1968 he travelled to Macau, China, to study traditional Chinese medicine for four years. On returning to the US, he put his newfound skills into setting up an acupuncture and herbal medicine clinic in Boston. His work was incredibly effective, and it was quickly noticed by the Western medical establishment; in the eighties and nineties he held positions at pain-research units in Boston hospitals and in 1998 was recruited by Harvard to research complementary medicine.[35] But Kaptchuck had realized that the power of his practice did not lie in the acupuncture or herbal medicines per se, but instead in the strong belief patients had in him: the placebo effect. So this is where he directed his research, with astounding (and somewhat controversial) results. A central tenet of the placebo effect is that for it to work, we have to believe we are receiving a real treatment. But, in 2010, Kaptchuck's lab carried out an experiment on IBS sufferers that turned this dogma on its head.[36] In the study, patients with IBS were randomly split into two groups. The first had a nice conversation with the doctor but were given no treatment. In the second group, the participants were given a placebo pill and told the following: 'Placebo pills made of an inert substance, like sugar pills, have been shown in clinical studies to produce significant improvement in IBS symptoms through mind–body self-healing processes.' Staggeringly, the patients in this 'open-label placebo' (also known as 'honest placebo') group showed much greater improvement than those in the no-treatment group. Placebos still work even if patients know that they are not real. Kaptchuck's team also recorded similar findings in patients suffering from other pain conditions, from chronic low back pain to migraine.[37,38] In the migraine trial, while these 'open-label' placebos were not as effective at relieving pain as the conventional treatment – rizatriptan – they were more effective than no treatment at all. In fact, the placebo was more than 50 per cent as effective as rizatriptan. One big takeaway from these findings is that, theoretically, placebos could be given ethically, or at least without deception. Perhaps this will only work when we are good at identifying who responds well to these open-label placebos. This may be achieved by working out whether genetic markers in individuals correspond to strong placebo responses – a brand-new area termed the 'placebome'.[39] But if it does work, we may have a pain reliever that has no side effects, and no risk of overdose or addiction.

Another way in which placebos could be used clinically is through the fact that the placebo effect can be subconsciously learned, or conditioned. It's clear that expectations around placebos are altered by verbal suggestion, but repeated, direct experience is another way in which expectations can be altered. When study participants were given a placebo alongside repeated applications of pain, the strength of which was surreptitiously reduced to make the participants believe the pain-relieving effects were stronger (known as preconditioning), the placebo effect was increased fivefold and this effect lasted for several days.[40] Associating a medication with a response eventually leads to the brain learning and adapting our expectations of future responses. This conditioning is not too unlike Pavlov and his dogs, which would subconsciously salivate at the sound of a bell they had learned to associate with food. In a 2016 study carried out at the University of Colorado, Boulder, participants who underwent preconditioning showed strong pain relief from a placebo cream on the skin, even though they knew it was a placebo.[41] Importantly, preconditioning does not have to emerge from our own past pain experiences; it can be shaped by observing and hearing others respond to pain treatments.[42] Tantalizing early evidence suggests that interspersing placebos with real pain-relieving drugs in a way that means the patient does not know which one is which – and informing the patient that this has been done – may reduce the need for the patient to increase their dose of the real pain-reliever, reducing side effects, dependence and cost.[43]

The fact that open-label placebos work is mind-blowing, and why this happens isn't yet fully known. Kaptchuck believes that conscious expectation or subconscious preconditioning are not sufficient to explain the bizarre phenomenon. For one, they do not explain the strong effect these honest placebos have in patients with a background of repeated medical failure and who show no signs of preconditioning. Put simply, our current biomedical model has revealed a lot of exciting scientific and practical information about the placebo effect, but it hasn't fully explained it. The open-label placebo throws a spanner in the works for anyone trying to find a unifying placebo theory. Over the decades of his research, however, Kaptchuck has been slowly coming round to what could be a unifying theory not only of the placebo effect, but of all neuroscience. When I first came across 'predictive processing', I had to make sure that I wasn't accidentally reading science fiction: it's counterintuitive and outlandish. But, while there's as much speculation about it as there is fact, a supportive body of evidence is steadily growing.

What if I told you that you can predict the future? And that you do it all the time? The conventional biomedical theory of perception is that the brain is a passive receiver of sensory inputs that come from both outside and inside the body (sight, sound, noxious signals, etc.), from which it then creates a perception, or model, of the world. This is a 'bottom-up' approach: you stub your toe, danger signals travel to your brain, which assesses the signals and produces pain. In the predictive processing model, the brain is constantly refining its idea of the outside world; it's balancing its own expectations, theories and beliefs (termed 'priors') of how the world should appear with new, incoming sensory inputs to form a best guess of what's going on in the outside world. The brain is a predictive machine. It's hard to see or imagine this in action, but what I if told you that the intentional error in this sentence wasn't an oversight by my editor? Your brain made you see what it expected you to see. Now have a look at the two upside-down heads below.

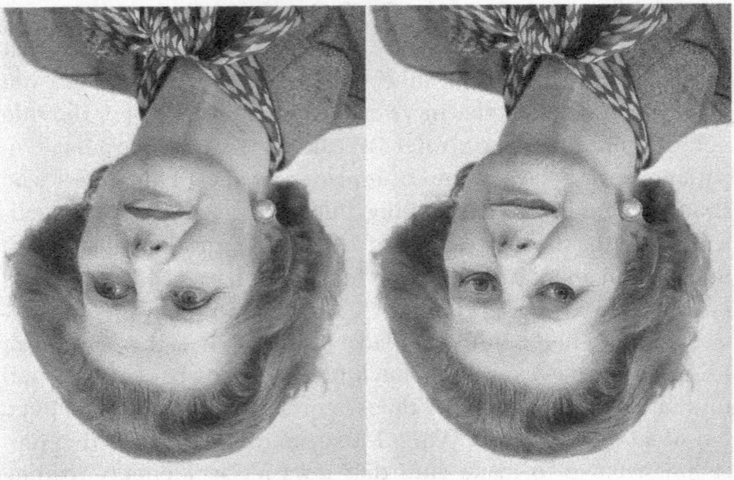

The 'Thatcher' illusion

Next, turn your book upside down and look at the heads again. It's now very apparent that in the image on the right the woman's mouth and eyes are upside down. This illusion – termed the 'Thatcher' illusion, as the first picture used for this trick was one of the late British prime minister[44] – shows that we perceive what we predict. All of these predictions are happening constantly, at incredible speeds, and this process is mostly unconscious. If you think about it from an adaptive, efficiency-related perspective, it makes sense for the brain to do this so that it doesn't waste energy on computing the entire overwhelming torrent of sensory information we are exposed to every second. This allows it to focus on what it is really interested in: 'prediction errors'. These errors occur when the incoming sensory information differs from what the brain expects. If they are very small errors, they're usually considered 'noise' and never reach areas of the brain involved with perception, so our depiction of the outside world doesn't change. If the error is large enough, however, our brain is forced to decide whether to revise its model of the world. Awareness of our surroundings only occurs when sensations violate our brain's expectations.

This data-filtering process is actually very similar to how a JPEG works, enabling the storage and sending of images without a perceptible loss of resolution. The value of any one pixel usually predicts that of its close neighbours, and the differences are found only along clear boundaries between objects in the image. The code can be compressed by coding only the unexpected data: the only thing transmitted is the prediction error. To use vision as an example: when light hits our retina, our brain has already created a prediction of what it thinks it will see; it is only errors in our brain's predictions that are passed on to higher levels of the brain.[45] This concept is consistent with the long-known anatomical fact that there are far more fibres travelling down from the visual cortex (the area of the brain that processes visual information) than travel up to it; this smaller signal only carries the *difference* between expectation and sensory information.[46] Otherwise, we see what we expect.

A stark example of this comes in the form of a viral internet phenomenon. Sometime in the spring of 2015, late to the pub, I could see a gaggle of my friends in the corner of the room, huddled around a smartphone and in heated argument. When I reached the group one of my friends grabbed the phone and held it under my nose imploringly: 'Come on, Monty, what colour is this dress?' I looked down at what was, obviously, a white-and-gold striped dress. My friend was visibly dismayed: 'No! It's clearly blue and black. What witchcraft is this?' The photo of 'the dress' appeared to some as blue and black and to others as white and gold. It divided friends, families and lovers, seemingly randomly. What was also fascinating is that most people couldn't easily change the way they saw it, as is possible in most optical illusions. A slew of scientific studies followed in the wake of the viral phenomenon, and while there's still no consensus, there's evidence to suggest that it's all down to predictive processing. When your brain is faced with ambiguity, it has to make a decision one way or the other as to what you will see, and this is likely based on past priors, assumptions about whether the dress is lit by artificial or natural light, or whether it is in shadow. My favourite theory is that our daily exposure to natural light or artificial light influences our brain's assumption when it sees the dress: early birds like me tend to see it in natural light (as white and gold) whereas night owls tend to see it as black and blue.[47]

What we see is not exactly what's out there in the world (although it's usually very close): it's our brain's best guess of the world, a guess made before we even see it. Professor Lars Muckli, a neuroscientist testing out the predictive processing model with highly advanced neural imaging, as part of the EU's Human Brain Project, puts it eloquently: 'Vision starts with an expectation of what is around the corner'.[48] Our prediction enables us to anticipate the movement and direction of an object before it has moved there: useful for tennis, life-saving when it comes to crossing the road. Our predictions also influence what we perceive. If you are trekking through a region of the Amazon rainforest known to be infested with venomous snakes, your brain will have a lower threshold for interpreting an ambiguously shaped branch as a dangerous reptile than if you were rambling through an English wood.

Perhaps this is also why we can't tickle ourselves: our brain can predict the exact sensations we will feel, as it knows the movements that the tickling hand is about to make, so the sensations are not exciting or surprising – our brain knows exactly where and when we will be touched. Throughout our lives, our brain gathers statistics to adjust and refine its internal model of our body and the outside world. According to predictive processing theory, these statistics are governed by Bayes' Law, a mathematical formula devised by Thomas Bayes, an eighteenth-century English church minister, to help calculate odds in games of chance. The formula calculates the probability of a prior being true, based on new evidence. When we are faced with prediction error, we update our beliefs in light of new evidence.

In a series of papers published in 2018 and 2019, Ted Kaptchuck combines current evidence with a dash of conjecture and argues that the experience of pain relief is not a direct consequence of healing in the body, but the process of the brain recognizing that healing is taking place or that the pain stimulus has been removed; the brain's hypothesis that we are in danger or damaged is being revised by a change in bottom-up information and cues.[49,50] If we have external cues that our pain is being relieved, our pain relief is quicker and stronger. This would explain why the participants in Irene Tracey's intravenous drip experiment only received mild pain relief when they were given a powerful opioid without their knowledge, but the relief was much greater when they were told that the infusion was running. Cues are not simply direct verbal suggestions: the therapeutic ritual of taking a pill, talking to a calming and confident doctor and being in a clinical environment prompts the brain into thinking that any changes within the body are causing pain relief, and to minimize the prediction error, the brain reduces the perception of pain.

Kaptchuck believes that this theory explains the mystery of the open-label placebo effect. When I took the homeopathy pill – a placebo – in the depths of my IBS flare-up, there was dissonance in my mind: should my brain change the prediction of continued pain because of the cues and ritual of the pill-taking behaviour, or should the prediction continue because I know, logically, that the pill is inert? Kaptchuck argues that, in these times of mismatch, our non-rational, automatic and emotional responses can override our conscious, rational mind. This open-label placebo effect may be due to neurological events being unconsciously influenced by the ritual performance of healing, and not by conscious thought. Kaptchuck maintains that 'placebo effects are primarily elicited by what you do, and only secondarily – or not at all – by what you think'. It's important to reiterate that this is still speculation, and open-label placebo science is in its early infancy. But the predictive processing model also neatly explains why expectation – conscious or unconscious – influences perception, and why pain is amazingly malleable. This model also goes beyond the placebo effect when it comes to explaining pain; why it is so easily influenced by concentration, emotions, expectations and past experiences as well as tissue damage. Perhaps this could help people conceptualize and deal with persistent pain. Imagine someone has long-term low back pain and their priors include a fear of moving or loadbearing because they were told – years ago – not to further aggravate their 'slipped disc'. In the vast majority of cases of long-term back pain, gentle, graded movement is actually good for pain relief, so if a well-informed physiotherapist, osteopath or doctor explained this to them, and slowly helped them to bend and lift, they'd likely feel pain but not as much as they expected and feared. A large prediction error would have been created that gradually changes their internal model of the world: they start to associate movement with safety and reduced pain, and begin a journey of recovery.

Our brain's expectation and prediction has enormous influence on how we perceive pain. This is a square peg that doesn't fit in the round hole of Western medicine's mind–body dualism. It will be a while before we know whether or not giving 'honest', open-label placebos for pain will work, but there are so many ways in which we can all use 'the expectation effect' to reduce pain and improve lives. Cultivating a positive treatment context – from the physical environment of a clinic room to the demeanour of the doctor – is not just an added bonus: it's essential. Anyone privileged to be in a position of providing care to someone in pain must strive to secure trust, stop unnecessary anxiety-producing language, strengthen positive associations and create a context of realistic but strong positivity. Harnessing the expectation effect is not about giving out sugar pills or encouraging treatments that work only via placebo, but about fostering informed confidence and reducing anxiety. Most importantly, it makes medicine human, pointing both the person in pain and the treatment-giver in the direction of hope and healing.

5

The Meaning of Pain

The power of emotion and psychology

The oldest and strongest emotion of mankind is fear, and the oldest and strongest kind of fear is fear of the unknown.

H. P. LOVECRAFT

For our physiology to calm down, to heal and grow, we need a visceral feeling of safety.

BESSEL VAN DER KOLK

EVAN IS AN affable and articulate Australian, able to seamlessly move from extolling the benefits of beer-temperature controllers to explaining the nitty-gritty of international humanitarian law. As we chatted, I found it almost impossible to imagine that this wise head behind a cheeky grin had been through hell on earth.

In 2006 Evan had achieved a lifelong dream: the twenty-three-year-old soldier could now wear the 'sandy beret'. One hundred and sixty tough, fit and intelligent soldiers had undergone Australia's Special Air Service Regiment (SASR) selection process, and only nineteen – of which Evan was one – had passed. He was now a member of an elite special forces unit with an international reputation. And they were preparing to see action. That year, the Australian government announced the deployment of a mentoring and reconstruction task force to Uruzgan Province in Afghanistan. The task force's role was literal and metaphorical bridge-building, while the role of the special forces was to identify, detain or neutralize threats to the task force. After the special forces captured enemy combatants, they would be detained by Australian interrogators from the Defence Intelligence Organisation (DIO) for a period of no longer than ninety-six hours before being handed over to Afghan security forces. In preparation for this task, the DIO interrogators practised their techniques on young SAS soldiers. With instructors of limited experience and no accountability, it was a recipe for disaster.

'"Resistance to Interrogation" training is designed to expose you in a limited and controlled way to what happens if you are captured,' Evan told me. 'It's not designed to be torture, it's not designed to break you, it's not even designed to take you near that point. It's meant to be exposure.' As part of their training, the soldiers were meant to hold out for at least forty-eight hours while only providing the 'Big Four': name, number, rank and date of birth. In the articles of the Geneva Convention, a prisoner of war is bound only to give this information.[1]

Evan was bagged, dumped into the back of a truck and driven to an undisclosed interrogation centre. 'Once you got to the centre, they took your equipment and tried to recreate who you are, such as what patrol you belong to. They pride themselves on being able to get all this information – and they're good at it. But I knew all this beforehand, and had sterilized every piece of equipment and removed anything that could give away my role or what patrol I belonged to. This was clearly a challenge for them, and it was clear that at the outset they decided "right, we're going to break him".'

Evan was tortured for almost one hundred hours. 'It's so incredible what they're able to do to you . . . how malleable they can make you. It's as though you're under a spell. You're deprived of all your senses. You're wearing blacked-out ski goggles. Blaring music is coming at you from all directions. You're handcuffed and made to sit – completely naked, except for a hospital gown – on a cold cement floor with your legs splayed out in a stress position.' Evan would later find that these techniques had been directly taken from Guantanamo Bay, where, in the years following 9/11, interrogators became experts in causing excruciating pain without the incriminating evidence of tissue damage.

Evan was taken out of the room nine times over the course of the ninety-six hours for interrogation. The interrogator would sit directly opposite him and shout the words: 'NAME. NUMBER. RANK. DATE OF BIRTH. NAME. NUMBER. RANK. DATE OF BIRTH. NAME ...' The persistent, hypnotic demand is a strategy to lull prisoners into giving away more than they should. But Evan didn't yield. After one interrogation, he requested to be escorted to the toilet. Denied this, he was instead viciously assaulted by the instructors. After he was dragged back to his cell and placed in a stress position on the cold cement floor, Evan noticed blood running down his legs. One of his assaulters had kicked him so hard in the buttocks that they had caused an anal tear. The slow hours of torture continued.

Evan had been driven to a state of helplessness and humiliation and was under constant physical threat. He was in genuine fear for his life. The lightest touch or subtlest muscle-twinge now caused pain: 'even the anticipation of going through pain again caused its own pain'. Psychology, emotion and context are just as important to the experience of pain as physical input – and indeed are often more so. 'There's pain – sometimes extreme pain – you can tolerate. Take "weekend warriors", for example, who do triathlons and crazy ultradistance races. It's a controlled atmosphere: you're in control, there's no real threat and you have a goal. You can endure that pain with a goal in mind. And when I was going through SAS selection, it was frequently painful, but I had the goal of the sandy beret and we knew that we could quit at any moment. But this was very different.' Roughly seventy-two hours into the torture, during which he also had neither food nor sleep, Evan lost consciousness.

In the first months following this ordeal, Evan wasn't able to express or even understand what had happened to him. One evening he walked into the toilets of his barracks and suddenly began to hear a disembodied voice echoing around the bathroom, repeating the words 'NAME. NUMBER. RANK. DATE OF BIRTH . . .' This was the first of many symptoms of post-traumatic stress disorder (PTSD). One of the worst of Evan's symptoms, and one that confused and distressed him immensely, was an extreme sensitivity to pain. He prided himself on a high pain threshold and an ability to tolerate the gruelling labour required of an SAS soldier, but now he felt agonizing pain when putting on boots or entering a swimming pool of any temperature cooler than a warm bath. The pain was often felt over his whole body. Evan's traumatic experience of torture had forced his brain to make a rapid jump from short-term to persistent pain. His brain had essentially been rewired to be on hyper-alert for any potential threat. If pain can be crudely likened to an outdoor security light that switches on when a human-sized intruder approaches your house at night, Evan's light had become one that is triggered by a leaf blowing in the wind. Persistent pain is very common after torture but is under-recognized as a post-traumatic symptom. Interestingly and crucially, the development of long-term pain after torture is dependent not on the extent of physical damage sustained during the experience, but on the psychological and emotional impact of the torture and the emergence of PTSD.[2]

Pain is both sensory and emotional. These elements overlap and intertwine in terms of both the brain's physical real estate and our own lived experience. They are often so blended as to seem indistinguishable. Scientists have long known this; the most widely accepted international definition of pain describes it as 'an unpleasant sensory and emotional experience . . .'[3] Torturers, too, have long known this, and are experts in how our emotions and thoughts influence pain. Evan's interrogators built a sense of anticipation and threat, removed any sense of control he had over his own body, humiliated him and subjected him to pain at unpredictable times in seemingly random ways.

But the same emotional circuitry that torturers exploit can be used to help people deal with, and sometimes eliminate, pain. Before delving into this new and exciting research, we have to briefly define our terms, which isn't helped by the fact that there is no scientific consensus on the definition of emotion, and scientists still hotly debate it.[4] But, by and large, most will agree that emotional experiences are feelings that are outcomes of biological activity in the body. Each emotional experience we feel is unique, although it usually falls into a distinct category: fear, anger, disgust, etc. Picture an emotional experience as a cake emerging from a concoction of ingredients: inputs from the peripheral nervous system (a stubbed toe or an empty stomach), cognitive processes (memories and attention), decisions and mental evaluations. There are different (and usually recognizable) types of cake, but each one is slightly different and unique in its own way. Emotions can be evoked by physical sensations (e.g. that empty stomach), environmental events (e.g. witnessing a car accident) or deeper cognitive processes (remembering a time you made a fool of yourself).

When it comes to blending emotions with sensory inputs to create a unified experience, one of the head bakers (to eke the most out of the analogy), is an area of the brain called the anterior cingulate cortex (ACC). This boomerang-shaped area of the brain sits snugly between the 'emotional' limbic area of the brain and the 'cognitive' prefrontal cortex, and this anatomy is crucial for its role. The ACC monitors streams of sensory information coming to the brain from the body, such as noxious signals from a stubbed toe. The ACC is constantly on the lookout for damage and danger in the body, but considers itself more a professor of pain than a mere detector. In its ivory tower in the frontal cortex, the ACC doesn't waste time with trivial pain information, such as where noxious signals are coming from, but instead looks for the *meaning* of pain. It integrates physical, emotional and social elements of pain in experiences such as social exclusion, anxiety and depression. The ACC has helped us understand that when someone 'hurts our feelings' we are actually hurt and literally feel pain. The ACC is activated not just when we feel physical pain, but also when we feel the hurt of rejection.[5] Remarkably, the painkiller paracetamol reduces the hurt of emotional pain and social rejection, with imaging studies showing that it dampens down activity in this region.[6] Studies have shown that over-the-counter 'physical' pain medications also reduce emotional responses to evocative images and reduce discomfort when people part with possessions.[7] The importance of the ACC and other areas of the emotional-pain circuitry is unambiguously recognized when a stroke or brain tumour selectively damages them. Anna, the Austrian woman with pain asymbolia in Chapter 2, knew when she was experiencing pain but the unpleasant, emotional quality of it was completely lost. As pain was not aversive, it didn't protect her from unnoticed injury. Without its emotional element, you could argue that what Anna felt was no longer really pain.

This emotional anatomy of pain has recently been exploited by a group of maverick neurosurgeons. Deep brain stimulation (DBS) is a neurosurgical procedure that can be used to relieve a number of conditions, in which tiny pacemaker-like electrodes working at specific frequencies are carefully placed in specific areas of the brain. It was first used to relieve some types of pain in the 1950s and is still used today, often as a 'last-chance saloon' for various types of intractable chronic pain, including post-stroke pain – pain caused by stroke-damage to the areas of the brain that process pain.[8] In the procedure, electrodes are placed in brain regions involved in sensory pathways – areas that handle the sensation and discrimination of pain. But Andre Machado, a neurosurgeon at the Cleveland Clinic, Ohio, was not happy with the mixed success of DBS, and decided to go off-piste. At the 2017 meeting of the American Association of Neurological Surgeons,

Machado and his team presented some astonishing findings.[9] In their study, neurosurgeons performed DBS and implanted electrodes in the brains of patients with chronic post-stroke pain. But instead of placing electrodes in areas associated with sensory pain pathways, Machado placed them in areas of the emotional brain, specifically the ventral striatum and the anterior limb of the internal capsule. The results were unexpected but fascinating. Overall, patients did not report a reduction in their pain intensity; if it felt like a nine out of ten before the surgery, it still felt that intense afterwards. But there was a dramatic improvement in their mood, sense of well-being, independence and quality of life. While pain intensity remained unchanged, suffering was relieved. The meaning and experience of pain had changed. This game-changing study could open up avenues for relieving the most severe, stubborn pain by modulating areas of the brain related to emotion.

You may be thinking that the emotional element of pain is restricted to the extremes of torture and the cutting edge of neurosurgery. But all the pain we all experience, whether short- or long-term, is moulded by our moods, emotions and thoughts. Professor Irene Tracey, at the University of Oxford, has played with a number of emotions to see how they shape the pain experience in healthy volunteers and in people living with persistent pain. Among the first emotions she explored – and perhaps the most powerful – were anxiety and its close sibling, fear. In a 2001 study, Tracey had healthy male volunteers lie in an fMRI scanner.[10] Each volunteer was given a visual cue in the form of a shape, such as a triangle or a square. Roughly ten seconds after this cue they would receive a blast of heat to the back of their left hand. One symbol (e.g. the triangle) was always followed by a moderate heat stimulus; another symbol (e.g. the square) was initially followed by the same moderate heat stimulus but, as the experiment continued, the appearance of the square would sometimes be followed by a much hotter stimulus. To the participants, the square slowly became the dreaded square. As the participant became uncertain as to whether the square was going to be followed by an intense heat, their anxiety levels rocketed. Even if the moderate pain stimulus followed the square, it would be perceived as more painful than the exact-same stimulus when it followed the triangle. On the fMRI images, one particular area of the brain, the entorhinal cortex, was consistently activated in the anticipation of pain and caused increased activation in areas of the brain associated with the emotional processing (the anterior cingulate cortex) and intensity-coding (insula) of pain. This well-designed study shows that anxiety worsens pain, and the fear–pain relationship can become a self-fulfilling prophecy.

Take my fear of needles, for example. As a doctor, taking blood from patients and giving injections is as second nature as brushing my teeth. However, when it comes to visiting the doctor myself, and being on the receiving end of such needles, my fear of them has not eased since childhood. The anticipation starts on entering the surgery and builds in the waiting room, and even the neutral tone of 'Dr Lyman to Room Three, please' gets my pulse and mind racing. This concoction of emotion and overthinking means that a small pinprick, which most other people barely notice, feels as though I'm being skewered by a jouster's lance. It may be the fear that the doctor or nurse giving the injection may just happen to be one of my more incompetent friends from medical school. Or maybe a particularly painful injection experience in the past led to severe anxiety before my next injection, which itself worsened the experience of pain, which worsened anxiety leading up to the injection after that, which worsened the subsequent pain. Ad infinitum.

On an individual level, my vicious cycle of needle fear and pain is utterly trivial compared with Evan's experience. But on a grand, population-wide scale it is immensely important. Let's look at vaccines. They work; they are one of the greatest medical interventions we have ever created. For each birth cohort (group of people born in the same year) in the USA, the childhood immunization programme prevents roughly 20 million cases of disease and 40,000 deaths.[11] And vaccines have been the only long-term solution to eliminating the scourge of Covid-19. But the needle-fear–pain cycle directly interferes with the work of vaccines and other medical interventions; children who go on to develop needle fear are likely to avoid not only vaccinations, but also blood tests, dental care and donating blood.[12] They will then be less likely to vaccinate their own children, will pass on the fear, and the cycle continues. Non-compliance with vaccinations both endangers the individual and lowers herd immunity, putting others at risk. Importantly, the fear–pain cycle also causes misery to the individual, as it increases anxiety and pain with each vaccination. Worse still, vaccination pain is generally under-prioritized by doctors: it's seen as just as a 'sharp scratch' – that phrase most healthcare professionals use, even if it doesn't reflect the actual experience – and pain is assumed to be an accepted part of the procedure. The good news is that there are lots of practical, low-tech and evidence-based ways of reducing needle pain, setting people up for long-term positive relationships with future vaccinations and invasive procedures.[13]

The idea is not to pretend that the injection won't hurt – in fact, telling a patient that it won't hurt could even worsen the experience of pain by creating an environment of uncertainty and dishonesty. Instead, the goal is to increase feelings of safety and comfort and to minimize feelings of threat and danger. With infants, the parent or caregiver should sit them upright and give them a cuddle; children shouldn't lie down – this increases fear because it reduces the child's sense of control, worsening pain. If infants are young enough, the vaccination should be given while they are breastfeeding, or they can be given something tasty and sweet, like sugar water. These simultaneously provide the pain-relieving elements of distraction, pleasure and safety. Enjoyable distractions are vital: not only do games and jokes divert the child's attention; they help them associate the experience with happiness and safety. Deep breathing also induces a sense of calm and control. Getting the child to blow bubbles combines this with elements of entertainment. In children over the age of four, there is evidence that rubbing the area around the injection site before and during the vaccination reduces pain. Other evidence-based tips include using numbing creams beforehand and, if multiple vaccines are being given in one sitting, giving the most painful injection last.

Words are also incredibly powerful. Perhaps counterintuitively, reassuring the child with phrases like 'Don't worry, it'll soon be over' and 'Aw, I'm sorry about this' actually makes the child think that there is something to worry about. Bear in mind that the subconscious doesn't really listen to the words 'not' or 'don't'. These are just prefixes; the brain latches onto the object, the word that has meaning. Instead, the key is for the caregiver to talk about the positives after the vaccination – what went well – and to encourage the child; this reduces anxiety and the likelihood of pain the next time they have a vaccination. In cases of severe fear and anxiety, psychological therapies such as exposure-based therapy are often effective and definitely worth trying. With the individual and societal effects of the fear–pain cycle in needle-phobia, it's not just a sharp scratch.

Anxiety and fear worsen pain because they increase the sense of damage, danger or threat, which makes the brain want to protect the body even more. The areas of the brain associated with emotional pain (such as the ACC, insula and prefrontal cortex) also play an important role in deciding whether the pain is self-induced or caused by something external. A research team in Beijing created a slightly twisted torture instrument to demonstrate this.[14] They turned grip-strengthener rings into inverted knuckle-dusters by attaching beads with pointy ends to the inside of each ring. The subjects would then have the ring wrapped around their left hand and would either squeeze that hand with their right hand or have it squeezed by the examiner. The same pressure was deemed much more unpleasant when the examiner squeezed, and imaging confirmed that the brain was interpreting the pressure induced by the external threat differently.

This concept of an external threat, coupled with a loss of control, explains why Evan's torture was so excruciatingly painful. Strapped down to a cold, dank floor for hours, with sight removed by goggles and hearing overwhelmed with a volley of loud music, his brain interpreted any touch as a potentially damaging, life-threatening encounter and his pain system went into overdrive. The pain-worsening effect of helplessness has been known for a long time. A 1948 experiment subjected rats to electric shocks whenever they ate.[15] One group of rats learned that they could temporarily stop the shocks if they jumped up, whereas the other group received shocks regardless of whether they jumped up or not. The latter group appeared significantly more disturbed and were unwilling to eat. Two decades later, the American psychologist Kenneth Bowers found that human subjects given a sense of control over electric shocks – being told that they could and should avoid the shocks – rated them as much less painful than those who were told there was nothing they could do to avoid the shocks.[16] Evan was clear that the pain he experienced in his SASR selection process, in which the soldiers were told they could stop as soon as they signalled, was deemed much more bearable than the pain during his torture ordeal, over which he felt that he had no control.* This is not just of relevance to the realm of torture (experimental or real). If people living with persistent pain are

* Evan's formal complaint about his treatment resulted in a seven-year fight with the Australian Special Forces and a high-profile court case during which he garnered the support of MP Jacqui Lambie, who used parliamentary privilege to bring his case into the public domain. Evan, who had been removed from his unit and stripped of his rank, eventually had his rank reinstated and his extensive legal costs paid for by the Department of Defence.

given a sense of control and empowerment, the intensity and unpleasantness of the pain itself should diminish. Clearly, the best way of doing this is by explaining what pain is and what it isn't. The next best route to empowerment is to give people techniques with which to deal with their pain day by day.

Anxiety is not the only emotion that can significantly worsen short-term pain. If you are in a positive mood but want to see how quickly your happiness can be taken down a few notches, listen to Sergei Prokofiev's orchestral piece 'Russia Under the Mongolian Yoke' at half speed. This work has been successfully used to induce sadness in numerous psychological studies.[17] Irene Tracey's team had healthy student volunteers listen to this piece while reading negative statements such as 'My life is a failure' and 'I have no friends'.[18] As if these participants weren't unlucky enough, they would receive a pain-inducing burst of heat from a stimulus applied to their left forearm. They would later receive exactly the same pain stimulus while listening to more cheerful music – the Largo from Dvořák's 'New World Symphony', and reading neutral statements. The students reported the pain to be more unpleasant in the sad condition than the neutral condition. This is probably unsurprising, but confirms an important point: our mood influences our experience of pain. Importantly, the study demonstrated the physiological processes behind this by imaging the participants' brains in an fMRI scanner during the tests. When a sad mood was aroused, there was increased activity in a number of brain regions responsible for both the sensory and emotional aspects of pain, including the amygdala, insula, inferior frontal gyrus and anterior cingulate. Tracey showed that manipulating someone's emotional state in a negative way turns up an 'anxiety volume button' in the brain. This makes sense in light of what the experience of pain boils down to: it is essentially all about protecting yourself from danger or threat. If we are anxious, or fearful, our brain is going to want to amplify this alarm signal. If pain is fire, these emotions are petrol.

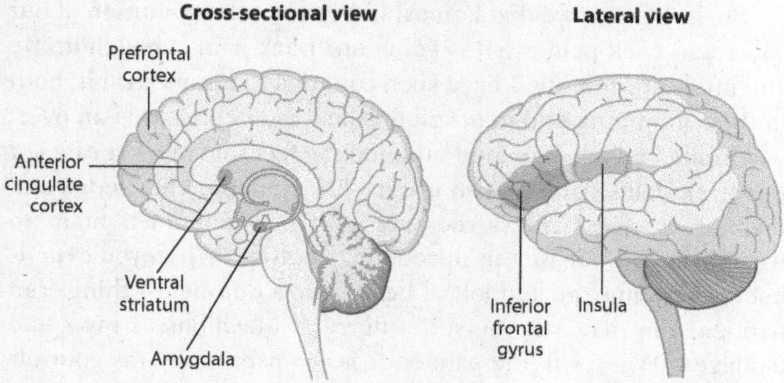

Emotional processing of pain

Negative emotions and feelings of threat do not just worsen short-term pain; they can facilitate the transition from short-term to long-term pain and they can help wire the brain to expect pain and suffering. Let's take long-term back pain, for example. It's incredibly common in the West; it's the largest cause of work-related absence and something the majority of us experience at some point in our lives.[19] When our back hurts, it is distressing and it is completely understandable to assume that we've done some serious damage. But what's really curious is that there is a very weak connection between back pain and the condition of our spine. A huge body of research shows that most cases of chronic back pain show no signs of structural abnormalities, and many people with these abnormalities (such as disc herniation) don't have any pain at all. In fact, in the vast majority – more than 90 per cent of cases – there is no identifiable tissue damage. Also, there are fewer cases of chronic back pain in developing areas of the world – including among people who undertake 'back-breaking' work in their daily lives and have no ergonomically designed chairs or specialized mattresses.

So, if there's a weak relationship between the condition of our spine and back pain, what is going on? Back pain is real, horrible and often life-ruining. I have seen countless patients with it, both in the community and in hospital. The good news is that an overwhelming body of evidence supports the fact that in most cases of persistent back pain, we can use readily available knowledge and techniques that harness the power of our emotional brain to weaken and eliminate this miserable condition. When you experience a new, unexpected jolt of back pain, a number of things can happen. You may ask yourself a flurry of questions: is my spine damaged? When will the pain end? Is the pain under my control? The pain, particularly if it persists, then activates the areas of our frontal cortex involved in rumination and catastrophizing. The predominant emotion is fear of damage and – as pain is ultimately about protection – this greatly amplifies the pain. One small, temporary muscle-twinge in your lower back urges you to protect your back, but can easily convince you that your spine is damaged. This fear could lead to hypervigilance – our brain interpreting usually painless signals as reasons to assume danger and damage – and lead you into a self-perpetuating vicious cycle of fear and avoidance of any kind of movement, when staying active is actually one of the greatest balms for chronic pain.[20] In fact, it may not be long before your brain associates any kind of movement with pain. Your avoidance starts to occur in anticipation of pain, instead of in response to it. Feelings of anxiety, threat and low mood worsen the pain, which in turn worsens those moods. These negative moods insidiously rob you of other natural painkillers: good sleep, socializing and a healthy diet. They also distort your body's hormonal and immune systems, amplifying chronic stress and worsening pain. Down and down the vicious spiral you fall.

This transition has been documented across a number of imaging studies. One 2013 study at Northwestern University, Illinois, analysed a group of patients who had short-term back pain and followed their progress over the course of a year.[21] At the end of the year, the difference in the brain scans between those whose pain disappeared and those who developed chronic pain was fascinating. When back pain shifted from short-term to long-term pain – even when the cause of the initial pain had gone away – the neurosignature of brain activity shifted towards emotion-related circuitry in the brain. As pain becomes long-term, it engages more of the brain associated with emotions and fear (the amygdala, prefrontal cortex and basal ganglia). It appears that this signature can become deeply ingrained in the brain, as they found similar results in those who have lived with persistent back pain for at least ten years. This transition from acute to chronic pain is, perhaps unsurprisingly, more likely if someone has an underlying mood disorder.[22] People who worry more about their pain are also at higher risk of prolonged pain and opioid dependence.[23]

Fortunately, we can break this cycle. Our brain remains wonderfully adaptable – or 'neuroplastic' – throughout our lives. Through reframing and retraining, the fear–pain cycle can be broken and pain can be significantly diminished, even completely relieved. One approach is through talking therapies, or 'psychotherapies' as they are more formally known. A 2016 study carried out at the University of Washington found that two types of psychological therapy were more effective at relieving chronic low back pain than 'usual care', which involved giving the participant $50 to seek whatever treatment or medication they would usually take.[24] One was cognitive behavioural therapy (CBT) for pain. This both educates people about pain and gives them techniques to recognize and change negative thoughts that influence the pain experience. The other was mindfulness-based stress reduction (MBSR). This involved training in mindfulness meditation and yoga, encouraging people to give attention to thoughts, emotions and sensations without trying to change them.

Another psychotherapy garnering a lot of attention is acceptance and commitment therapy (ACT). The word 'acceptance' is understandably contentious when introduced to people living with terrible, long-term pain. 'Are you really expecting me to just roll over and accept my pain?' I remember one patient saying. But acceptance is not about giving up and giving in; it is actually a very effective starting block towards gaining control over one's pain and developing psychological flexibility. Among many other things, ACT is about accepting your immediate experience of pain – being a non-judgemental observer – instead of trying to control or defeat it. This is often helped through mindfulness, and over time reduces ruminations – persistent negative thoughts – and strong emotional reactions to pain. ACT enables many to live with their pain, and it can even reduce or sometimes eliminate it.[25]

A good (albeit extreme) place to look for insights into how mindfulness-based psychological therapy works is in expert meditators. A team at the University of Wisconsin–Madison did just this, subjecting Buddhist meditators who each had at least 10,000 hours of meditation under their belt to laser-induced heat pain.[26] Interestingly, the pros experienced equal pain intensity to novices, but reported markedly reduced pain unpleasantness. Imaging revealed that in the experts there was an initial increase in activity in emotional-pain-related regions (the anterior cingulate cortex and insula), which reduced with repeated stimulation and also resulted in low activity of the amygdala (which plays a crucial role in triggering fear and anxiety) prior to pain. The study's authors suggest that cultivating 'experiential openness' to pain reduces the anticipation of pain and its sense of threat, and the anxiety and fear that come with that.

Persistent pain is often worsened – and sometimes caused – by past traumatic experiences, and addressing these can sometimes powerfully relieve people's pain. A novel psychotherapy, designed by teams at Wayne State University, Detroit, and the University of Michigan, teaches people to view their pain as being strongly influenced by emotions, relationships and past trauma, and helps them express both positive and negative feelings, thus becoming more aware of their own emotions. Recent evidence shows promise that this emotional awareness and expression therapy (EAET) is effective at reducing persistent pain.[27] Evan ultimately found that psychological relief, not strong medication, was the only way he was ever going to experience pain relief. Evan's excruciating, long-term, whole-body pain was almost completely eliminated by a psychological therapy targeted at those suffering from past trauma: eye movement desensitization and reprocessing (EMDR). This evidence-based treatment for post-traumatic stress disorder (PTSD) involves recalling traumatic memories while performing a bilateral stimulation task, such as following someone's finger moving from one side to another with your eyes. The general idea is that the patient can only retrieve a limited amount of information about the trauma while performing the task, so the negative emotions associated with that trauma become less intense. The memory is reprocessed in a less negative way, and the patient becomes desensitized to it. This list of psychological therapies is not exhaustive, and there are various other therapies that have been developed,[28] but the effective ones have some things in common: they empower the individual by educating them about pain, they reduce the sense of danger and threat, and they facilitate the healthy processing of emotions.

I am not suggesting that a mind-over-matter approach will be effective for everyone and all types of pain. Indeed, a large study published in August 2020 found that – on average – psychotherapies such as CBT and ACT, while often helping people to deal with pain, have only a small effect on reducing pain on their own.[29] These specific therapies are likely very effective for some people, but make no difference in others. It's also important to be aware that it is dangerous and wrong to view and treat pain as if it were just a mood or emotion. And psychotherapy can easily backfire: treating someone's real, terrible pain as though it is 'all in your head', or a cognitive distortion that can just be wished away, is scientifically incorrect and, frankly, offensive to the person in pain. All that said, mood, emotion and psychological outlook have an enormous influence on pain, and we gloss over them at our peril. Changing the way we perceive pain and the emotions associated with it requires time and a lot of practice, but it is low-tech, has few side effects and can transform lives.

Cultivating a positive outlook and knowing that things will get better, even if it requires hard work and patience, can reduce the suffering and disability caused by pain, and even relieve it altogether. Another key takeaway is that anything that improves mood when mood is low – whether it's socializing, engaging in purposeful activities, or antidepressants – is also likely to improve the experience of pain. What's more important than improving mood for mood's sake is educating the individual so that they can reframe their outlook from one of fear and hopelessness to one of confidence and hope. Healing from long-term pain is about moving from an outlook of threat and danger to one of protection and safety; the transition from seeing pain as an informant of tissue damage to seeing pain as a guardian angel that wants to protect us but is often overprotective. Our mind is a powerful modulator of pain, and meaning is a powerful medicine.

6

No Pain, No Gain

Pain, pleasure and purpose

I don't mind pain, so long as it doesn't hurt.

ANON. (ATTRIBUTED TO OSCAR WILDE)

Content warning: The last section of this chapter contains descriptions of self-harm.

I LEARNED AT SCHOOL that the influential eighteenth-century English philosopher Jeremy Bentham, the father of 'utilitarianism', had asserted: 'Nature has placed mankind under the governance of two sovereign masters, pain and pleasure.'[1] The ultimate goal of human well-being, utilitarians argued, is to maximize pleasure and to minimize pain. It seems simple: pain is the only evil; pleasure is the only good. However, a controversy-courting classmate of mine took exception to this and – being always one for a wisecrack – raised his hand: 'But, Sir, but what if I like a good spanking?' Cue a bashful teacher trying to divert a group of sixteen-year-olds away from the subject of sadomasochism. But the quip had insights beyond its intent.

I had learned from experience that pain and pleasure are not always opposites, and even that pain can be pleasurable. A month or so earlier, I had enjoyed one of the most precious, hallowed days of any British child: a snow day. School was off, so I joined a group of local friends on the hunt for an impromptu snowball fight. I cannot remember the identity of the rival tribe – most probably pupils from another school – but when each snowball struck my body and face I felt a surge of purpose-filled elation, not pain: I was taking a hit for my friends, showing them (and myself) my commitment to the cause. The following day, while the snow hadn't completely melted but school had reopened, I was walking down the road, books in hand, when I felt a sudden, punching pain in the small of my back immediately followed by a puff of ice blowing over the back of my neck as the snowball disintegrated. I spun around to see my younger brother emerging from a bush with another snowball ready to throw at me. Objectively, the strike on my back was softer than most of the hits I had taken the previous day, but it was much more painful and a shroud of discomfort settled over my back for the next hour; the context had changed – the game was over and the attack unexpected.

This anecdotal example of similar sensations being painful in one context and pleasurable in another has long been suspected but has been demonstrated in a controlled lab setting only relatively recently. A particularly dramatic example of one identical stimulus being painful in one situation and pleasurable in another was seen in a 2012 study by Irene Tracey's team at Oxford University, using the first experiment to show 'hedonic flipping'.[2] Two groups of participants were exposed to a heat stimulation on their skin. Each member of the first group, the control, had periods when they received a warm, non-painful stimulus and other periods when they received a moderate-intensity heat stimulus, intended to cause mild pain. The second group was exposed to the moderate-intensity stimulus and a high-intensity stimulus designed to cause more pain. After each exposure, the participants were asked to rate the intensity of the stimulus, recording how painful or pleasurable it was.

In the first group, the moderate heat stimulus was interpreted as unpleasant and painful. In the second group, however, the moderate stimulus came as a relief, as it was much better relative to the intense pain-stimulus. Surprisingly, for many in the second group the moderate pain-stimulus was often interpreted as pleasurable. These reported experiences were also corroborated by biological findings. Neuroimaging using fMRI scanning confirmed that those experiencing 'pleasant pain' had reduced activity in areas of the emotional brain (insula and anterior cingulate) and even had increased activity in areas associated with reward circuitry (the prefrontal and orbitofrontal cortices), dampening down pain signals travelling up the spine. This reward system is activated by the best outcome in a given situation, such as gaining money or experiencing a small loss of money when large losses are possible. Remarkably, this reward system is relative, dependent on context, so the study participants found relief – even pleasure – in moderate pain stimulation in the context of intense pain being the other option.

This clever study showed that the exact-same stimulus can cause pain in one context and bring relief and pleasure in another. By changing the context in which pain is experienced, it is possible to switch the experience of pain from negative to positive, termed the 'hedonic flip'. This seems to go against everything we know about pain: pain is meant to hurt; pain is an experience that motivates us to change our behaviour, usually in an aversive way, to avoid whatever is causing the pain. But, fascinatingly, if one type of pain is the least-painful option in a certain situation, our brain makes it pleasurable, even encouraging us to seek it.

Even after reading this intriguing study, I assumed that it was something of an academic anomaly; my education and experience continued to tell me that, by and large, humans seek pleasure and avoid pain. This is clearly crucial for survival, as is tragically seen in those with congenital analgesia, who cannot feel (and thus do not avoid) pain, and rarely survive past their teens. As far back as the fourth century BC, Aristotle could see that pain and pleasure governed our behaviour; that humans 'choose what is pleasant and avoid what is painful'.[3] He, too, saw this as a good thing: 'in educating the young we steer them by the rudders of pleasure and pain'. Jeremy Bentham would be happy with all of this: pain equals bad, pleasure equals good. But then I slowly stepped back and looked at the real world and actual human behaviour; if you measure our activity by any number of different means – the volume of spicy chillies eaten, the total distance of miles jogged, the number of copies of *Fifty Shades of Grey* read – it is evidently clear that humans often actively seek pain.

The key to understanding this is in the role of pain as our protector and, as we seek protection, as a motivator of our behaviour. We humans behave as we do to thrive and survive, by seeking rewards and avoiding punishments. The human body is also constantly treading a tightrope, trying to achieve internal balance and equilibrium. Any stimulus that pushes our body towards that equilibrium is felt as pleasurable; anything that pushes the body away from stability is felt as unpleasant. Pressing your forehead against a bag of frozen peas might be intensely pleasurable on a sweltering summer's day, but it would be miserable on a cold winter's night. A slice of stale bread is haute cuisine to someone who hasn't eaten for two days, but revolting to someone who is full. In a similar way, pleasure is a sign that a stimulus is rewarding or useful to someone; pain is a sign of danger or punishment. The nearer a stimulus brings the body to equilibrium, the greater the pleasure and reward value of that stimulus.

Our interpretation of pain or pleasure is also dependent on our knowledge of future rewards or threats. The greater the sense of danger or threat, the greater the unpleasantness of the pain. When people experience pain that's not associated with a threat to health or life – such as labour pain or controlled pain in a research setting – they often rate the degree of unpleasantness as lower than the intensity of the pain; but with cancer or chronic pain, patients give the degree of unpleasantness of pain a higher rating than its intensity.[4] Equally, the expectation that you are going to have a reward in the form of pain relief in the near future delivers real pain relief itself – a core tenet of the placebo (or, should I say, expectation) effect.[5] Importantly, if a reward is on offer but can be achieved only at the price of a small amount of pain, we are often happy to go through with the pain. The experience of pain can even enhance the pleasure of the reward. A simple, fundamental example is the toil and pain humans have gone through to grow and hunt food.

The shifting and often flipping relationship between pleasure and pain becomes even more mercurial when you throw social and cultural influences into the pot. My younger brother is an officer in the British Army and has – to my mind, at least – become conditioned to enjoy pain. This metamorphosis from innocent humanities graduate to hardened warrior began with a year of 5 a.m. starts and gruelling exercises at a military academy, and was completed at the legendary Platoon Commanders' Battle Course, which seems to consist solely of three months' running up Welsh mountains in the rain, carrying bricks. Whenever we catch up, he teasingly sneers at my body – one that could happily jog a few kilometres but would not survive a week in the army – and improvises a garden boot camp. As I struggle through my thirtieth press-up, to make my suffering more humiliating he gleefully screams a medley of painful platitudes: 'Pain is weakness leaving the body!', 'Sore today, strong tomorrow!' and, of course, 'No pain, no gain!'

While these motivational quotes are said with a sadistic irony known only to younger brothers, they say something profound about pain, pleasure and purpose. While studying at the Queen Elizabeth Hospital Birmingham, where the Royal Centre for Defence Medicine is based, I often came across young soldiers who saw pain as necessary and purposeful, even pleasurable. A psychological study at the centre found that there were two differing approaches to pain exhibited by service personnel, which changed depending on context. These are the 'no pain, no gain' view – pain being necessary – and the 'roughie-toughie' image, where feeling or exhibiting pain is seen as a weakness.[6] Social learning and cultural expectations influence pain behaviour and even make it pleasurable or desirable, whether it is a boxer conforming to the tough, hardened image expected of a fighting athlete or those raised in an honour–shame culture, where not expressing pain – even in the throes of illness – is a source of pride (as we will see in Chapter 9). Whenever I assess a patient in pain, I find it immensely helpful to get to know their background and fixed beliefs about the purpose of pain.

Human behaviour clearly shows that we do not see all pleasure as good and all pain as bad. Instead of my initial assumption that we seek pleasure and avoid pain, we actually – and this is a subtle but crucial distinction – seek rewards and avoid punishment. These rewards enhance our body's equilibrium, social acceptance, and, ultimately, our protection and survival. And what constitutes a reward can vary greatly within and between individuals. When we experience pain or pleasure, the melange of physical sensory inputs, our body's internal balance and our awareness of potential rewards and threats combine to create what is termed the 'subjective utility' of someone's pain: the meaning of pain. This takes us to the heart of one of pain's core conundrums: why there is so much variability between the intensity of a stimulus and the final experience of pain. The mechanisms behind this disparity are neatly explained in the 'motivation–decision model' of pain, put forward by Howard Fields, Professor of Neurology at the University of California San Francisco, who claims: 'the variability in the relation between the strength of a noxious stimulus and the intensity of the resulting pain experience can be understood as the manifestation of a decision process'.[7]

This decision is between whether we respond to a pain-causing stimulus or focus on a competing drive. Importantly, these decisions are not made with our conscious awareness. Within a fraction of a second, these competing interests have been discussed and weighed up by the Ministry of Defence in our brain, which makes a decree to either produce the experience of pain or not. An extreme example of this would be someone who is bitten in the leg by a large, uncontrolled dog while jogging in the park and does not feel the pain at all because they are focused on a much more important drive: fighting – and then either destroying or escaping – a direct threat to life. Ultimately, when faced with a motivational conflict, anything that is immediately more important to survival than pain will produce analgesic effects. This also helps a woman in labour to manage her pain as she brings her child into the world. In this model, it is also important to note that pain can be endured as long as the reward has pain-relieving effects.

The underlying neuroscience behind the decision-making in these pain–pleasure conundrums is complex, but it can be largely boiled down to two key characters: opioids and dopamine. Although morphine and heroin are certainly the most famous opioids, many are produced within our own body and are necessary for liking a stimulus or experience – for feeling good. For example, increasing opioid signalling increases the pleasure from food rewards, whereas blocking opioids markedly reduces the pleasure from food.[8] Importantly, blocking opioids also reduces reward-related pain relief.[9] While opioids are necessary for liking something, dopamine is the neurotransmitter necessary for wanting something. Dopamine acts before we obtain the reward. It encourages us to take action, either to seek a reward or avoid a punishment. Dopamine plays a key role in reducing current pain when a future or imminent reward is expected. Humans given dopamine have a significantly enhanced expectation of pleasure from future life-events.

In a study at University College London, subjects were asked to rate their expectations of happiness if they were to go on holiday to each of a list of eighty destinations around the world. They were then all given a placebo and a list of half of the destinations and told to imagine holidaying in these places. Then half of the group were given dopamine (in the form of the Parkinson's disease drug L-dopa) and the other half a second placebo, and all were asked to imagine holidaying at the remaining destinations. The next day, participants had to pick one destination from pairs for which they had given equal ratings the day before, and then they rated all of the eighty destinations again. Remarkably, expectations of pleasure significantly increased for the destinations imagined under the influence of dopamine.[10] Dopamine encourages us to seek rewards, but, importantly, it is also released when a painful stimulus is stopped, demonstrating that the relief of pain is also perceived as a reward.[11] When it comes to the pleasure associated with relief from pain, it actually pays to be a pessimist. A study conducted at the University of Oxford found that pessimistic participants experienced a greater sense of pleasure when pain was relieved than optimists did.[12] This is because the reward from relief comes from violation of a negative expectations and the emotional element of surprise, both of which are larger in those not expecting a good outcome.

To see how dopamine and opioids work to make pain pleasurable, let's take a look at Katie, who is approaching mile twenty of the London Marathon. A few hours of running has produced large amounts of lactic acid in her muscles, which irritates nociceptors that in turn fire noxious signals towards her brain.

But as she has an expectation of the significant personal and social reward of finishing the race, dopamine is produced in a phased release from a part of the brain called the ventral tegmental area. This causes the release of opioids – such as endorphins – from structures crucial to the reward circuitry of the brain: the nucleus accumbens, ventral pallidum and amygdala. These areas also release opioids when experiencing pleasure itself, not just in its anticipation. This release of opioids ultimately inhibits the noxious signals coming from Katie's legs, blocking signals that would usually result in pain. One reasonably common genetic mutation in humans produces high levels of phased-release dopamine and a subsequently high release of opioids during exposure to painful stimuli.[13]

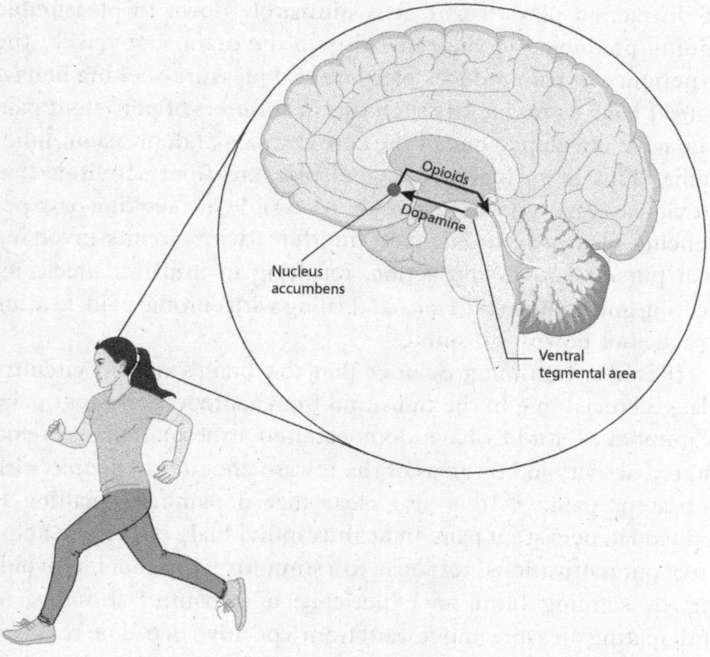

Reward circuitry and pain relief

Fascinatingly, there is also a remarkable overlap between the pain and pleasure centres in the brain, particularly in those areas involved in the brain's reward system. One window into the interlinked nature of pleasure and pain is seen in an Ohio State University study, which found that the pain-relieving drug paracetamol not only improved the emotional element of pain – making unpleasant stimuli feel less unpleasant – but also reduced feelings of pleasure.[14] It's also clear that pain and pleasure directly affect each other. It has long been evident that pleasurable experiences such as food, sex or music markedly reduce pain.[15,16] While an element of distraction plays a part, it is ultimately down to pleasurable stimuli producing feelings of safety in the brain. Conversely, the experience of pain reduces pleasure and pleasure-seeking behaviours. I have so frequently seen this in sufferers of persistent pain who have developed one of the core features of depression: anhedonia. This is an inability to feel pleasure from activities that previously gave a person great joy. Alongside not seeking or experiencing pleasure, chronic pain interferes with circuits involving goal pursuit and taking action, resulting in impaired decision-making and coping strategies in dealing with chronic pain, leading to a vicious downward spiral.[17]

It is also becoming evident that the brain's reward circuitry plays a crucial role in the transition from acute to persistent pain. A number of studies have demonstrated structural changes and altered activity in key areas of the reward circuitry in people with persistent pain.[18,19] It is also clear that dopamine-signalling is reduced in persistent pain. In healthy individuals, dopamine helps drive our motivational response to a stimulus – whether it is avoiding, or learning from, an experience of a painful stimulus, or anticipating pleasure and reward from a positive one. The reduced dopamine-signalling in persistent pain results in reduced motivation and feelings of depression.[20] In patients with fibromyalgia – a disease of persistent, widespread pain – the ventral tegmental area (which releases dopamine) has significantly reduced responses during both the anticipation of pain and its relief compared to healthy controls.[21] This may help explain the heightened pain sensitivity in sufferers of fibromyalgia and other persistent pain conditions. Understanding the complex relationship between pain, pleasure and the brain's reward system will be crucial in finding better pharmacological and psychological therapies for the scourge of persistent pain.

It's clear that humans can actively seek some level of pain if it brings about a reward, and if the brain decides that there is something more important for surviving and thriving than pain, it will exert analgesic effects. This is all well and good; as Katie runs the London Marathon, her pain is relieved – and she may even experience a runner's high due to her brain pumping out opioids and cannabinoids, our natural pain relievers – as a result of the perceived reward of finishing the race. But why do we get a kick out of pain for pain's sake? Eating a hot chilli pepper doesn't seem particularly rewarding. If you are one of the hundreds of millions of people across the globe who love the burning sensation of chilli peppers on your tongue, you are a partaker in 'benign masochism'. This term was created by Paul Rozin, a professor of psychology at the University of Pennsylvania, who describes benign masochism as: 'Enjoying initially negative experiences that the body (brain) falsely interprets as threatening. This realization that the body has been fooled, and that there is no real danger, leads to pleasure derived from "mind over body" '.[22]

All humans, to an extent, seek activities that cause a degree of pain in order to experience pleasure, whether this is found in spicy food, vigorous massages, masochistic sexual practices or stepping into a too-cold or too-hot bath. The key is that it is a 'safe threat'. The brain perceives the stimulus to be pain-provoking but ultimately non-threatening. Intriguingly, this could be similar to the way humour works: a 'safe threat' that causes pleasure by playfully violating norms.[23] We feel uncomfortable, but safe. In this context, where survival is clearly not at stake, the desire for pain is actually the desire for a reward, not suffering or punishment. This reward-like effect comes from the feeling of mastery over the pain. The closer you look at your chilli-eating habit, the more remarkable it seems. When the active ingredient of chillies – capsaicin – touches the tongue, it stimulates exactly the same receptor that is activated when any of these tissues are burned. Knowing that our body is firing off danger signals, but that we are actually completely safe, produces pleasure. All children start off hating chilli, but many learn to derive pleasure from it through repeated exposure and knowing that they will never experience any real harm. Intriguingly, seeking pain for pain's sake appears to be uniquely human. The only way scientists have trained animals to have a preference for chilli or to self-harm is to have the pain always directly associated with a pleasurable reward.

The idea that the act of intentionally harming oneself – causing physical tissue damage as well as pain – would lead to any sense of relief is difficult to understand for most people who have never self-harmed. In fact, when my childhood friend Ellie (not her real name) confided in me that she had been cutting herself for two years, even she didn't seem to know why she was doing it. The day after her fourteenth birthday – on which she'd had a storming argument with her then boyfriend – she'd sat alone in the family kitchen while her parents were out at work and begun to rub a bundle of wire wool over a small area of skin on her forearm. She continued to scrape back and forth until she had broken through the skin. 'When I felt that first trickle of blood across my arm, I felt a massive sense of calm; of peace in myself. Then, obviously, I felt really guilty and tried to hide the scar from my parents and friends. I don't know why I did it and I didn't want to do it ever again – but I became consumed by it, I guess.'

This behaviour is sometimes termed 'negative reinforcement', which is when you do something that removes bad feelings or emotions. Ellie was angry with her boyfriend and had a deep, gnawing anxiety that she would never have a stable or satisfying romantic relationship. She says that every time she cut her skin, once the physical pain subsided it seemed to 'pull out, exorcize the worries from my body', albeit for a short time. It didn't feel more complicated than that; it wasn't a form of self-punishment and Ellie says she definitely didn't want to bring any attention to herself. Meanwhile, for some people, self-harm can be a means of 'positive reinforcement', in which doing something (in this case, causing pain) gives a positive reward – for example, someone with depression wanting to feel alive in a sea of emotional numbness. Less commonly, self-harming can be a means of communication – as a cry to others to help or to stop doing something. Sadly, I have come across many people, including doctors, who believe that self-harm is at best attention-seeking and at worst a form of manipulation; two reasons that make up a tiny minority of cases.

While it seems evident that self-harm relieves negative emotions, a group of researchers at Harvard University took on the challenge of working out the mechanism behind this elusive, sad and seemingly contradictory phenomenon. In a set of studies between 2010 and 2013, researcher Joseph Franklin found that when self-injurers were subjected to pain during his experiments (including placing their hands into ice-cold water and giving them electric shocks), the relief when the pain was stopped made them feel even better than they did before they were subjected to the pain.[24] Remarkably, the control group (people who do not self-harm) also experienced this post-pain euphoria. This is termed 'pain-offset relief'. This is why, when Ellie's physical pain subsided, her emotional pain – which uses much of the same neurological real estate in the brain – subsided too. The team at Harvard also found that, over time, if a self-injurer uses the same stimulus (such as a razor blade) to cause the pain and its subsequent pleasurable state of relief, the user associates the stimulus with pain relief and this reduces the unpleasantness of the pain itself. This could explain why so many keep coming back to cutting.

But while Ellie said that she always felt pain, and that its subsequent relief was necessary for the emotional release, many who self-harm do not feel much pain at all. In 2014, the team at Harvard found that self-injurers have higher pain endurance than those who don't self-harm, being able to hold their hands in ice-cold water for longer.[25] Interestingly, it seems as though those with the most difficulties regulating their emotions could cope with physical pain for longer, as could those with high levels of self-criticism. In terms of self-criticism, it appears the lower someone's self-esteem and the more they believe they are 'deserving of punishment', the longer they are willing to endure pain.[26] A 2019 study also found that, for those with very low self-belief (in both self-injurers and control groups) the experience of pain itself – not just pain relief – improves mood.[27]

When Ellie disclosed her self-harming to me, she also said that she had recently been diagnosed with emotionally unstable personality disorder (EUPD), also known as borderline personality disorder (BPD), a condition often characterized by difficulties in emotion regulation and associated with frequent self-harm. Studies find that, compared to control groups, patients with EUPD have increased activation of emotional brain circuitry (particularly in the amygdala, that almond-shaped nucleus in the brain crucial for emotional responses to stimuli) when shown pictures evocative of both positive and negative emotions. But when they are subjected to painful stimuli, the activation of the amygdala is suppressed.[28] Physical pain paradoxically inhibits emotional regions of the brain, bringing about a short-term reduction in emotional distress.

Jill Hooley and Joseph Franklin, two of the key Harvard researchers, used their findings to create a 'benefits and barriers' theory in which both the pain experienced in self-injury and its subsequent relief bring benefits: a reduction in negative mood and an increase in positive mood – both negative and positive reinforcement.[29] But there are also barriers to self-harm, perhaps the most significant being the aversion to the sights and stimuli of self-injury – blood, wounds, knives, razors and other sharp objects. An intriguing 2020 pilot study by Hooley's team used fMRI scanning to image the brains of self-injurers and healthy control groups while they were shown images of razors and injured wrists. The healthy controls experienced significant activation of the amygdala – showing normal fear and repulsion responses – but these were diminished in self-injurers.[30] Intriguingly, self-injurers also had increased activation in an area of the brain's reward circuitry when viewing these images. Although this is a pilot study and more evidence will be needed over the coming years, it may well prove that, over time, the brain of a self-injurer becomes rewired to change the meaning of pain (both the experience and the stimuli that cause it) from something aversive to something rewarding. The malleable meaning of pain in the context of self-harm is not just an academic curiosity; it gives us hope for helping people to rewire their brain to break free of negative, destructive cycles. Take, for example, the discovery that the increased pain-endurance of self-injurers is often due to low self-worth. By giving self-injurers targeted psychological therapy that specifically aims to promote self-worth, Hooley found that their pain endurance significantly reduced, moving towards a healthy drive to escape the pain: 'the more valuable that people feel, the less willing they are to endure a bad situation'.[31]

Pain and pleasure are not the 'two sovereign masters' we spend our lives avoiding and seeking respectively, as appears at first glance and as Jeremy Bentham thought. They are instead two servants used by our brain to enable us to seek reward and avoid punishment – and ultimately to survive. These servants are flexible and multitalented, and illustrate the brain's remarkable ability to make subconscious decisions in our best interests. They also demonstrate the crucial importance of emotions and thoughts in determining how we physically feel, and how we then behave. A painful stimulus can feel unbearable if coupled with threat, uncertainty or fear, but can be immensely pleasurable if in the context of safety, sexual excitement or the anticipation of a reward. Suffering can be rewarding and enjoyable if it conveys a meaning to the sufferer that helps them survive and thrive in their context and their society. If there's one thing that our complex pain–pleasure paradox should teach us, it is that factors often sneered at or ignored in the medical world – emotions, thoughts, social influences – are central to conditions such as chronic pain, addiction and self-harm, and should be central to their treatment.

7

I Feel Your Pain

Why pain is contagious

You never really understand a person until you consider things from his point of view . . . until you climb into his skin and walk around in it.

HARPER LEE, *TO KILL A MOCKINGBIRD*

'CAN YOU FEEL this?' I asked.

'Yep,' replied Joel almost wearily, the ever-so-slightly-knowing rise of his eyebrows revealing that everyone who meets him does exactly the same thing. 'It feels less real than if you were sitting in the room with me, but I can still feel it. I feel it on my left-hand side, like I'm a mirror.'

I had just stroked the right-hand side of my face and a man three thousand miles away had felt me touch him. I was video-calling Joel Salinas, a Harvard-trained neurologist who has a rare condition called mirror-touch synaesthesia. He feels a touch on his body when he sees it happening to someone else – or at least how his brain imagines the touch to feel. Synaesthesia comes from the Greek 'perceived together' and is the phenomenon where the brain processes different senses at once. The stimulation of one sensory input (vision, for example) results in the activation of an unstimulated sense (such as touch). Some synaesthetes associate letters of the alphabet with particular colours. Some attribute personality traits and genders to numbers. Some can even taste sounds. This intriguing phenomenon comes in many shapes and sizes; there are roughly seventy different recorded types of synaesthesia.

Joel's mirror-touch synaesthesia combines his senses of vision and touch. If he sees a baby's head being patted in the street or a pair of strangers embracing, he can bathe in the soft, comforting touches himself. But if he sees a needle piercing the skin of someone's arm, he feels that brief but raw, penetrating pain. So I asked him why, out of all the possible careers out there, he chose to become a doctor.

'Fair point! I didn't realize I was different to anyone else when I applied to medical school. Seeing patients in pain, or being cut open, is often difficult. But actually, I think that being able to feel a patient's pain helps me to empathize with them, and sometimes feeling their pain gives me subtle clues towards the diagnosis. But maybe I couldn't have been a trauma surgeon!'

He chuckled and I joined in. When I felt his smile reciprocated in mine, and recalled the times people unconsciously mimic the facial expressions or gestures of others, it began to dawn on me that maybe, to some degree, all humans are mirrors. While Joel lives at the extremities of human perception, perhaps he'd have something to tell us about ordinary pain experiences.

Mesmerized by chatting to the closest thing to a superpower-wielding comic-book hero I'd ever met, I continued my barrage of questions. 'You never have, and never will, experience childbirth. What happens when you see a pregnant woman with labour pains?'

'It's really weird,' he told me. 'But I do feel sensations in my abdomen. It's something I have to work through as I know that I'm not experiencing labour. Sometimes I just step back in my mind and enjoy the curiosity of the sensation.'

He explained that the intensity of the pain is stronger, the closer the person in pain resembles him.

'What happens when you see someone in hospital die?'

Joel said he certainly noticed that he reacted differently from the other medical students when he saw his first patient death. He puts it vividly in his book *Mirror Touch: A Memoir of Synesthesia and the Secret Life of the Brain*: 'As the doctors continued chest compressions, I felt my back pressed firmly against the linoleum floor, my limp body buckling under each compression, my chest swelling with each artificial breath squeezed into me through a tube, a hollow, sipping sensation. I was dying, but I was not.'[1] He then rushed to the hospital bathroom, throwing up into the sink, and desperately tried to reassure himself that he was not dead.

Developing mechanisms to create boundaries between himself and other people was a necessity. 'Over the years I've found mindfulness a really successful way to not get lost in the patient's pain, so to speak.' Some mirror-touch synaesthetes are rendered housebound – the only escape from becoming overwhelmed by the sensations of others. Amanda, a mirror-touch synaesthete interviewed on America's National Public Radio *Invisibilia* podcast in 2015, has no dining table in her house because she can't eat with other people, as she would feel each of their mouthfuls being forced down her own throat.[2] All the blinds in her house are permanently closed to shut out the overwhelming stimulation of other people; simply seeing them would burden her with their sensory world. Leaving the house is fraught with danger. One day, on a trip to the grocery store, she saw a young boy playing in the park slip backwards and hit his head. She recalled that, as she started to run to help him, 'all of a sudden my eyes went blurry . . . I was down on my knees before I could get to this kid . . . my head hurt so bad that I was crawling to him.'

Then I asked Joel a question that had to be asked: 'How do you diagnose or verify mirror-touch synaesthesia? The cynic would wonder whether you were making it all up.'

'That's reasonable, and it's so hard to objectively measure mirror-touch synaesthesia if you don't experience it yourself. It's like trying to measure empathy, and there's also so much variability along the synaesthesia spectrum.'

Joel did get the closest thing to objective verification, though, when he travelled to the lab of researchers Jamie Ward and Michael Banissy at University College London, where he was subjected to a barrage of tests. This included the 'visual–tactile congruity task', in which he had tappers (essentially small pins with plastic attachments) taped to his left and right cheeks, which were connected by wires to a computer. He then watched a video of a woman being touched on either her left or right cheek, or both. Each time her face was tapped, one (or both) of the tappers on his own cheeks would tap. After each tap, he would then press a button to indicate whether the real tap had occurred on his own left cheek, right cheek, or both. Joel found it incredibly difficult, often impossible, to tell the difference between the taps on the woman's face and the real taps on his face: both felt equally real. His results were strewn with errors: 'These clearly indicate that you are a mirror-touch synaesthete,' the experimenter told him.

What most intrigued me about Joel's remarkable trait was his claim that literally feeling others' pain has given him 'hyper-empathy'. If we define empathy as the ability to feel and understand someone else's pain, it does seem logical that Joel would have this in bucketloads, and there is recent evidence to support this. Ward and Banissy, the scientists who assessed him in London, have been working with mirror-touch synaesthetes for over a decade, and in a 2018 study they found that they do indeed have higher levels of empathy – by some measures – compared with the general population. They are better at identifying facial expressions of emotion and have higher levels of emotional reactivity to others.[3] Perhaps surprisingly, they do not have higher-than-normal levels of what is termed 'cognitive empathy' – being able to imagine taking someone else's perspective; being in their shoes. But what makes them stand out is their ability to lose the boundaries between themselves and others, creating a heightened sense of raw 'emotional empathy'. They personify one of my favourite definitions of empathy, courtesy of American professor and personal development guru Brené Brown: 'feeling *with* people'.

I left my interview with Joel with two big questions. First, do all or most of us feel other people's pain to some extent, perhaps subconsciously? Second, could feeling others' physical pain be the cause of empathy? But, before we go any further: after defining what empathy *is* we also need to define what it *isn't*. It isn't acknowledging someone's pain – that's pity; going a step further, feeling sorry for someone's pain – that's sympathy; wanting to relieve someone's pain is compassion, ideally leading to positive actions.

Joel is not a complete rarity. Mirror-touch synaesthesia may be present in around 2 per cent of the population.[4] There is likely a genetic component to this: it seems that a predisposition for synaesthesia does run in families, but environmental and developmental changes (of which we are not yet fully aware) decide on the form the synaesthesia will take.[5] But, curiously, it can also be acquired after strokes or limb amputations. Around a third of amputees feel sensations in their phantom limb when they see someone else being touched. This is likely due to changes in the brain's connection between observed touch and felt touch.[6] But although the majority of us are clearly not synaesthetic, it only takes a little observation to realize that pain can be contagious. We physically wince when someone falls off their bike and strikes the hard surface of the road. We squirm at scenes of violence or torture in movies.

I carried out a YouTube experiment of my own to try to see this in action. Within a few minutes of searching, I had come across a 'reaction video', where you can view someone's reactions in real time as they themselves watch funny, shocking or downright weird videos. I was watching the actor and YouTuber Tyrone Magnus viewing a video of two women sitting by a garden table, engaged in a tense arm wrestle. While the arm wrestle begins evenly matched, one of the women seems to gain momentum, and slowly forces the straining right arm of her opponent towards the table. Suddenly there is a terrible crack and – mercifully – the video of the arm wrestlers is temporarily blacked out on my screen. But it isn't for Tyrone, who, as soon as he sees the woman's right arm break, screams, grimaces and immediately supports his right arm with his left hand, holding it in a protective, sling-like position as though he had broken it himself. Most of the time, and for most of us, this is an unpleasant emotional reaction that doesn't feel exactly like pain, but it's not uncommon for humans to feel both the sensory and emotional components of pain in themselves when seeing others in pain. A large body of evidence shows that when we see someone else experiencing pain, this activates many areas of our brain associated with the direct feeling of pain.[7] When Tyrone viewed the arm breaking, not only were pain signatures activated in his brain, but the area of his brain that maps and locates his right arm was also activated. This is known as 'neural resonance', essentially mirroring pain. This mirroring is certainly a real and common phenomenon, but whether or not this process is caused by specific 'mirror neuron' cells in the human brain is hotly debated and would lead us down a fascinating but ultimately inconsequential rabbit hole.[8]

While we can infer quite a bit from seeing an overlap in brain regions light up in the scanner, in 2015 a team from the University of California, Los Angeles (UCLA) went one step further to prove the link between pain and empathy.[9] In an elegant study, they first put 102 participants through an fMRI scanner – a large number for a study of this type – and found that giving patients a placebo pain reliever reduced both the intensity of the pain they would be subjected to in the experiment (an electrical shock to the back of the hand) and their pain empathy when they saw images of someone they had met earlier in the day also being subjected to shocks. Areas of the brain associated with empathy and pain were also reduced on the fMRI scans. If this wasn't interesting enough, their second step was the *pièce de résistance*: fifty participants were given naltrexone, a drug that blocks opioids, and not only did this block the placebo effect – with the participants feeling significant physical pain again – but their empathy for other people's pain also returned. This study shows that very similar processes are going on when we feel pain ourselves or see others in pain. In the words the UCLA team used in the very title of their research paper: 'empathy for pain is grounded in self pain.'

To understand the anatomy of neural resonance, and its implications for empathy, we need to explore how this emerges in the developing brain. As a child's brain develops, we begin to see the first, rudimentary signs of feeling someone else's pain. In 2008, a team at the University of Chicago found that when seven-year-olds see another human in pain, their periaqueductal grey (PAG) – crucial for pain perception – and motor cortex light up.[10] So when a seven-year-old sees their father miss the nail with his hammer and strike his own thumb, the child understands that he must be in pain and becomes aware of their own thumb or hand, maybe reflexively clenching their fist or even feeling pain themselves. This transmission of pain from the person who is suffering it to the observer is a primitive building block of empathy and the first appearance of neural resonance. But this activation of the PAG in young children is not enough for understanding the pained person's emotions and experience, nor is it adequate for understanding whether the pain was inflicted intentionally or not. This happens when the PAG becomes coupled to the developing thinking and decision-making centre of the prefrontal cortex and areas of the emotional brain such as the anterior cingulate cortex (ACC), amygdala and insula.[11]

Between late childhood and early teens, most children have developed a sense of empathy sophisticated enough not just to feel for someone they see experiencing physical pain, but for others' emotional turmoil, too. They then develop the ability to feel empathy for abstract groups of people they haven't seen before: the victims of a train crash, survivors of a tsunami, the oppressed citizens of an authoritarian regime. Developing the ability to 'catch' pain from others is crucially important for healthy interpersonal relationships. Adolescents with sociopathic traits (previously and still commonly called 'psychopaths') are less responsive in emotional areas of the brain when they see others in physical pain. In one study, sociopathic adolescents underwent fMRI scanning and were shown images of body parts exposed to painful stimuli, such as a finger slammed in a door.[12] When they were told to imagine the body part was their own, the emotional areas of the brain were clearly activated, but this activation was significantly reduced when told to imagine the body part belonged to someone else. Intriguingly, the level of reduced responsiveness was able to predict the severity of the sociopathic traits.

The centrality of one particular area of our brain to our empathy apparatus is particularly intriguing. The ACC is an error predictor and a conflict monitor: it flashes up when there is a discrepancy with something that is expected – for example, receiving an unexpected amount of reward (either too little or too much) from a particular outcome. In Chapter 5 we saw that the ACC is also a detector and evaluator of physical, emotional and social pain. This is a brain area focused on our self-interest, from judging whether we're getting what we deserve to whether we're being socially excluded or experiencing physical pain. So it is curious that the ACC is almost always involved when we experience empathy with another's pain.[13] A South Korean study in 2010 may have gone some way towards explaining this, and it doesn't shine an optimistic light on human altruism.[14] The study researchers found that fear can be learned and avoidance can be conditioned just by watching others experience physical pain. Importantly, ACC activation proved crucial to this learning, showing that something more is going on than just watching: the observer needs to feel the pain. We learn from our own experiences of pain, but it's also advantageous to learn from others' pain. I wonder whether my fear of heights can trace its way back to my first trip to the circus, when I saw a trapeze artist slip and fall – probably around ten metres – to the ground and be carried out writhing in agony on a wooden door-turned-stretcher; or whether my cautious approach to cooking stems from seeing my younger brother, aged six, nonchalantly drop a mackerel into a pan of boiling oil, only for the oil to splash his left arm and forehead, causing agonizing third-degree burns. Caring and looking after others we see in pain is certainly a reason for the contagious nature of pain, but a large part of it may well be directly due to self-interest. If we sample even a small taste of another's pain, we're more likely to try to avoid the situation or stimulus that caused their anguish. On weighing up the evidence, Robert Sapolsky, Professor of Biology and Neurology at Stanford, reaches this conclusion and encapsulates it well in his 2017 book *Behave*. He says: '*Feeling* someone else's pain can be more effective for learning than just *knowing* that they're in pain.'[15]

As is the case with our own pain experiences, feeling the pain of others is influenced by a melange of cognitive influences, from past experience and culturally shaped beliefs to our judgement of how the person got themselves into their painful situation. These factors affect how much our own pain system responds to seeing others in pain. A fascinating 2010 study found that subjects had more ACC activation when they witnessed clips of AIDS patients in pain who had been infected with HIV from a contaminated blood transfusion, compared with contracting it from drug use.[16] Thinking back to when, as a junior doctor, I would assess patients in A&E departments, perhaps it was easier to feel the pain of a facial laceration of someone who had been slashed in an unprovoked attack than someone who had fallen off a bus shelter after ten pints of strong lager. Depressingly, ACC activation is much greater if a subject sees someone from their racial in-group in pain than someone from an out-group.[17] Unsurprisingly, the effect is greater in those with stronger pre-existing racial biases.[18] This is particularly worrying because the less pain you catch from someone in an 'out-group' experiencing pain compared with your own (whether the divide be racial, social or otherwise), the less you are likely to help them. There is some good news for humanity, though: a study carried out by teams in Australia and Chile found that spending time with people of a different race increases the amount of pain felt when observing the pain of others from out-groups.[19]

Joel Salinas' mirror pain was stronger the closer the person suffering the pain resembled him. Although, for most of us, our sensitivity to others' pain is of a much lower intensity than Joel's, the evidence shows that this is the same for all of us: it is easier to empathize with those who behave and appear more like us. When the person in pain resembles us, it is easy for the emotional components of our pain pathway to generate a sense of empathy; when they are different from us, the cognitive areas of our brain (such as the prefrontal cortex) have to pick up the slack. In short, there is more thinking-effort required to imagine ourselves in their shoes. There are practical ways we can use this to our and others' advantage. Perhaps the simple act of admitting that we all have implicit biases, and then pondering these facts, will train areas of the prefrontal cortex to help us feel others' pain when we see it. In trying to help others, we also want to reduce the amount of work our brain has to do. The more we can relate to the person in pain, the less work our brain has to do and the easier it is to empathize. The international charity sector has learned that the most successful appeals and requests for donations do not focus necessarily on the number of people affected by disaster – from an East African refugee crisis to cyclones decimating areas of India – but instead describe real and relatable stories of the individuals affected. In the words of Mother Theresa: 'If I look at the mass, I will never act. If I look at the one, I will.'

Feeling someone's pain and developing empathy is, however, much trickier when the person suffering is someone you really don't like. When, as a Princeton PhD student, Mina Cikara attended a New York Yankees baseball game wearing a Boston

Red Sox hat, the hatred spewed towards her in songs and swearing inspired her thesis. With her then professor, Susan Fiske, Cikara conducted a number of experiments to assess intergroup envy and hatred, as well as *Schadenfreude* – the German word describing taking pleasure in someone else's suffering.[20] In one experiment, they monitored the brain activity of Boston Red Sox and New York Yankees fans in an fMRI scanner while showing them a series of baseball game excerpts on a screen. When the Red Sox and Yankees fans watched a neutral team – the Baltimore Orioles – playing, their brains appeared fairly uninterested as to whether this team succeeded or failed. But when they saw their rival team fail against the Baltimore Orioles, areas of their brain associated with reward and pleasure lit up on the fMRI scan: pure, unadulterated *Schadenfreude.* Other experiments – using self-report, scanning with fMRI and measuring smile strength (through the electrical activity of cheek muscles) – demonstrated that we are much more likely to feel good when misfortune befalls high-status, competitive targets – people we envy. Conversely, studies show that when people we envy succeed, our ACC is significantly activated – their gain is our pain.[21]

Clearly, overcoming this inherent feeling to look after a suffering rival would require a lot of prefrontal cortex energy. And it may be even harder for men. A study at University College London involved subjects playing a game together based on 'the prisoner's dilemma', a situation where participants can either help or double-cross one another.[22] Their brains were scanned as they saw other participants being zapped with electricity; empathy-associated areas were activated when they saw those who played fairly being subjected to pain. What the researchers perhaps did not expect was that male participants not only showed a significantly reduced empathetic response when seeing double-crossing players receiving pain, but – and this was not seen with women – areas involved with pleasure and rewards also lit up. Perhaps men have an innate lust for vengeance; perhaps the sex differences are more cultural than biological. Either way, the pain (or pleasure) we catch from others in pain is very dependent on the meaning we take from their suffering.

I once explained neural resonance – this ability of ours to catch and experience another's pain – to an old friend, a philosophy graduate, who made an interesting point: 'Could this be our brain's morality lobe? We don't want to harm others because it would essentially be doing harm to ourselves?' I was able to look up some recent evidence to find an answer. In 2017, a team at UCLA were thinking along similar lines and set out to see whether neural resonance influences moral decisions. The researchers recorded volunteers' brain activity via fMRI while showing them two videos – one of a needle piercing a hand, the other of a cotton swab gently touching a hand. A month later the participants were given ten moral dilemmas on the subject of causing harm, with a choice of two actions for each one. For example: you are residing in a small town in a war-torn country. Murderous soldiers have just arrived in the settlement and the townspeople are well hidden. But you are sitting next to a baby that is about to cry. You could fatally smother the baby to prevent its cry, and thereby save the other townspeople; or you could not kill the baby, and let it cry, but risk giving the locals' positions away, leading to their deaths. The choice is between causing harm and saving more people – 'maximizing the outcome' – or avoiding causing harm to the baby. The researchers proposed that 'people who show greater neural resonance with the pain of others should evince stronger tendencies to reject harm in moral dilemmas'. In other words, people who experience more pain when seeing others in pain are less likely to harm the baby, due to not wanting to experience their own pain. However, they found no correlation between activation of the neural resonance circuits and people's willingness to harm one person for the sake of the greater good. According to this study, my friend's hunch was wrong, and neural resonance is not the seat of our morality. But I think that this is good news for humankind;

it suggests that it is our genuine concern for others that governs moral decision-making, not simply a self-centred desire to feel less emotional pain ourselves.

Catching another's pain seems to be beneficial to us, to others and to society, but it needs to be limited. One study placed electrodes on the heads of doctors and non-doctors and monitored their brainwaves (known as electroencephalography, or EEG) when they viewed painful stimuli.[23] When the control – a non-doctor – saw videos of body parts pricked with a needle, they exhibited normal brainwave activity. The doctors, however, showed reduced emotional arousal. They had developed a thick skin, and seemed more detached from the person in pain. This may seem heartless – and, indeed, I have worked with some doctors who display a worrying lack of concern for their patients – but this learned response is perhaps important for reducing compassion fatigue, or mental fatigue in general, among medics. Not responding viscerally to every patient's pain should free up the cognitive real estate in the doctor's brain to use their knowledge and skill to help the patient.

This leads us to a crucial, practical juncture. The experience of feeling another's pain is good for learning to avoid pain ourselves and is likely to be important for developing healthy social relationships. But – and this is a big but – feeling someone's pain does not necessarily lead to doing something about someone else's pain. Empathy does not equal compassion. In fact, it may often have the opposite effect. If we feel empathy for someone's pain and feel virtuous about it, the resulting feeling of reward may give us a sense of completion without anything practical actually having been achieved. And if seeing someone else's pain gives us pain, anxiety and distress, perhaps we are more likely to tend to our own needs than help others. The key to solving this apparent paradox is to foster feeling someone's pain but with an element of detachment. When subjects are asked to focus on themselves – feeling the pain of someone else in pain – they are less likely to help them. This makes sense, as seeing pain encourages us to protect ourselves. But if subjects focus on being other-oriented (for example, having training that focuses on compassion, rather than empathy), they are more likely to translate their feelings of empathy into benevolent actions.[24,25] It's not a bad thing to feel agonized by someone else's suffering, whether they're right in front of you or live on the other side of the world, but it's what we do with that pain that counts – taking a step back, giving our brain some space for rest and thought, and figuring out how we are going to do good in response. And when we are tending to someone else's suffering, we should take heart: their perception of us as empathetic and kind actually helps to soothe their pain.[26,27] We are friends, spouses, nurses, doctors and carers, not machines. Our humanity, if used wisely and lovingly, is a powerful medicine.

The discovery that we mirror the pain of others also has very direct implications for those living with persistent pain. Whenever Pam, a sufferer of persistent neck pain, saw another person crane their neck to get a better view – such as her teenage son looking down at his smartphone or builders inspecting a roof from the ground – she would feel a sharp twinge at the base of her own neck. The fact that she didn't know why this was happening made her confused and distressed, which worsened the pain. I briefly explained neural resonance to her – that our brain mirrors others' actions, mostly unconsciously but sometimes even producing the feeling of pain, and that this is a protective mechanism to enable us to learn and avoid damage. In her case of persistent pain, however, this mirroring system was overprotective. Armed with this knowledge, she gradually perceived the movements of others not to be dangerous at all. Slowly but surely, these encounters would no longer trigger flare-ups of persistent pain. I'm convinced that we can harness neural resonance to help people out of the mire of long-term pain. One very exciting treatment for persistent pain that harnesses our mirroring abilities is 'graded motor imagery'. This enables the patient to gradually access and control their mirroring system so that seeing others experience pain doesn't provoke pain in themselves, 'sneaking under the pain radar' in the words of the graded motor imagery founders.[28] (More on this in Chapter 11.) When we're talking to others who have persistent pain, it's important to find out whether their pain is worsened by mirroring, as we need to be aware of the potential effects of our words and actions. Knowing that pain is contagious is a key to helping others in pain, but it is also crucial in understanding how to convert empathy – 'feeling *with* others' – into actively helping them.

8

Pulling Together

Social pain

How unhappy, how utterly alone, always he suffers the savagery of his illness with no one to care for him.

Sophocles, *Philoctetes*

'Social pain' sounds abstract, but we all know what it feels like. The pain of rejection – from being picked last for a school sports team to not receiving an expected wedding invitation – can be horrible. While this upsetting but universal aspect of human experience has been around since time immemorial, social media has magnified social pain in every way possible. A generation is growing up one smartphone's distance away from instant social judgement and rejection, where value is measured in 'likes' and love-hearts, and an unseen mob can turn against you at any moment.

In 2003, researchers at UCLA discovered that the pain of social exclusion – the pain of 'broken hearts' and 'hurt feelings' – isn't just psychological: it really does hurt. And, fittingly, they discovered this by creating a social video game.[1] PhD student Naomi Eisenberger and her professor, Matthew Lieberman, came up with the ingenious 'Cyberball'. One by one, UCLA undergraduates would lie in an fMRI brain scanner and begin to play the video game, in which they and two other players would stand in a triangle and throw a ball to each other. In the game, the participant is led to believe that the other virtual players are other subjects playing on other screens, but in reality they are completely computer-programmed. The players begin passing the ball to each other in what Lieberman admits is 'the most boring game you can imagine', but at some point in the game the two computer-generated players start passing only to each other. When this happens, the brain of the now-excluded participant reveals something very interesting. On the scans, rejection in the ball-throwing game causes elevated activity in the anterior cingulate cortex (ACC) – the region implicated in generating the perception of physical pain, and which blends sensations with emotions and cognitions to try to work out the meaning of pain. Eisenberger went on to carry out experiments showing that people with lower thresholds to physical pain reported greater distress at social rejection, and those who reported greater levels of social distress in the ball-throwing game reported stronger feelings of unpleasantness when exposed to physical pain-stimuli after the game.[2] Physical and social pain are intertwined at deep neurological levels, and while on the surface this seems a bit bizarre, it makes sense when we go back to the essence of pain. Pain is a protector; it's an unpleasant feeling that drives us to avoid danger and reach a state of safety. Our brain also perceives separation from others to be harmful and potentially threatening to our survival. Eisenberger's first paper references previous studies showing that hamster mothers with surgical damage to their ACC no longer try to keep their pups nearby, and baby squirrel monkeys with similar neural damage no longer cry when separated from their mothers.[3,4]

This has huge implications for understanding and treating pain. During my medical training, the 'bio-psycho-social' model of disease had been around for a few decades but was starting to gain traction in the world of medicine and medical research. And that's a good thing: all diseases, and particularly chronic pain, cannot be understood or treated without taking someone's psychological make-up and social environment into account. But '-social' felt somewhat tagged on at the end of the portmanteau, with most lecturers, medical students and doctors just paying it lip service. Doctors have to retain vast amounts of information, with which they make big decisions in small time-frames. Understandably, they find comfort in the laws and certainties of biological mechanisms, whereas a patient's social context is complex, seemingly subjective and often messy. But the reality is that the 'social' – from individual beliefs to social interactions to the very structure of society itself – is immensely important to the experience of pain.

What if I told you that there was a disease that can't be found in medical textbooks, yet is worse for one's physical health than smoking, causes depression and suicide, is contagious and affects a large and growing proportion of society? This condition is called loneliness. Social isolation ruins lives and prematurely ends them, and is one of the starkest examples of how a social condition can be so deeply intertwined with pain. Pain causes isolation, and isolation causes pain. Rats with persistent neuropathic pain (caused by damage to a nerve) begin to decrease their interactions with both familiar and unfamiliar rats. If their pain improves (by being given gabapentin, which is often effective in neuropathic pain), their socializing resumes.[5] We see this so often in humans: pain discourages mobility, increases fear (of both pain and social stigma), lowers mood and worsens fatigue, all of which keep the sufferer immobile and hidden from the outside world, and they gradually become more and more isolated. As each thread in the individual's social web is cut, their world continues to collapse in on itself. Social isolation is awful for mental and physical health; it's a risk factor for suicide and can be as detrimental to your health as smoking fifteen cigarettes a day.[6] Social isolation in and of itself also worsens persistent pain.[7] It's a terrible, vicious cycle. Humans are social, but we also need agency, individuality and a sense of control over our bodies and our lives. Persistent pain also robs individuals of this control, making sufferers particularly dependent on a support network.

But there is good news. A large body of (often entertaining) evidence shows that social bonding is a balm for almost all kinds of pain. In 2010, psychologists at the University of Oxford (in one of the most stereotypically Oxford studies ever) found that teamwork increases pain thresholds.[8] Twelve athletes were taken from the university's top two male rowing crews: the Blue Boat and Isis. Rowers were tested in two sessions during the same week, in which they rowed for forty-five minutes on an indoor rowing machine. In one session the individuals rowed by themselves; in the other session rowers were lined up next to each other in groups of eight and rowed in sync, simulating a virtual boat. After each session, the rowers had their non-dominant arm squeezed by a blood-pressure cuff to stimulate ischaemic pain (when blood supply to the tissues is restricted). Despite the rowers' work rate being very similar across the two tests, the athletes who were tested just after rowing in a group could tolerate about twice as much pain as when they rowed by themselves. It's clear, then, that working in a group diminishes the experience of pain. It's likely that when the rowers are working as a team – literally 'pulling together' – their brains attribute more meaning and purpose to the task and pump out more endorphins, the body's natural morphine. But there is probably something else at work. In rowing, success is dependent on each individual member of the group rowing exactly in sync, and the study's authors argue that it is this synchronous activity that enhances social bonding and increases endorphin release, dampening down pain.

But don't worry; you don't have to find seven other people willing to haul a fibreglass boat down a river with you to engage in life-enhancing synchronous activity. Music has played a crucial role in socializing and survival in every culture across the world at every point in history. In modern society we idolize music stars and watch talented amateur singers fight it out on TV to become the next big thing. Perhaps that leaves you feeling that you're not musical, and performance is better left to the 'talented few'. But evidence shows that only a tiny minority of people have no musical talent to speak of.[9] More importantly, an enormous body of research shows that engaging with music is beneficial for mind and body. The growth of community and church choirs across the UK in recent years – thanks to their increased media representation – has been a wonderful answer to our loneliness epidemic. Group singing benefits the brain and body, improves happiness and well-being, helps members form strong and meaningful social bonds and reduces chronic pain.

To see how choir-singing impacts people in pain, researchers at the University of Lancaster carried out in-depth interviews with people who had joined a community 'pain choir'.[10] Many of the interviewees described a significant reduction in pain during and after singing: 'It's better than all the tablets in the world,' said one member. The medicine of singing is also very cheap and portable, and participants are able to take their medicine home with them to self-manage their pain: 'I just think singing is good for you. I sing at home now,' said another. Group singing reduces pain in a number of ways. The happy atmosphere of purpose and pleasure takes attention away from the pain and produces endorphins. But in the long run, singing also begins to change the meaning of pain, gradually making it less unpleasant and diminishing its power, with one singer stating: 'I'm a survivor and I've adapted my life according to how I'm able to live it.' Really importantly, singing gave many of the interviewees a sense of purpose ('If I hadn't started eighteen months ago, I would still be sat on the couch now') that enabled them to gradually increase activity levels and participate in other pain-diminishing activities such as tai chi and swimming. Ultimately, the activity gave people hope. Whether it's harmonizing with a choir or belting out Abba in your living room, music is medicine.

There is another synchronous, social activity that I think we could all get on board with. In 2012, a team from the University of Oxford – from the same inventive psychology department that inflicted pain on the rowers – found that laughter may really be the best medicine.[11] In their first experiments, the researchers asked participants to watch either clips from comedy shows – ranging from *Mr Bean* to *The Simpsons* – or parts of a factual documentary. Given that evidence shows that humans do not easily laugh when watching even the funniest material by themselves and that we are thirty times more likely to laugh when we are in a group,[12] all participants watched the clips in a group setting. After watching the clips, all of the participants were exposed to painful stimuli – either a freezing cold cuff or a squeezing cuff wrapped around an arm. The team also tested participants in a 'real world' setting, having them watch live shows – either comedy or drama – at the Edinburgh Fringe festival before subjecting them to the painful stimuli. What's particularly interesting about the study is that laughter not only increased pain thresholds, but was a stronger cause of pain relief than mood: if a subject was in a positive mood but wasn't laughing, their pain threshold did not increase. Although they were unable to concretely prove it at the time – as they were not able to image the brains of the participants – the team hypothesized that, like synchronized rowing, the physical and muscular act of laughing dampens down pain by opening the brain's drug cabinet and releasing opioids. Five years after the study, researchers at the University of Turku in Finland (with the help of Professor Robin Dunbar, one of the key members of the Oxford team) used advanced neuroimaging techniques to confirm that this does indeed happen.[13]

Group singing, dancing, sport, music-making and the rituals of religion are examples of wonderfully affirming, rhythmic social bonding that, for many people, are better pain relievers than drugs. But social interaction doesn't have to be synchronous to be analgesic. When it comes to managing long-term pain, simple friendship can be stronger than morphine. Robin Dunbar's team in Oxford has also found that the larger someone's social network, the larger their tolerance to pain.[14] Social bonding binds endorphins to opioid receptors in our brain, which calms down pain. These findings also corroborate the strong links between social isolation, depression and chronic pain. And meaningful social interaction doesn't just reduce pain; fostering a good social network is beneficial for all aspects of mental and physical health.

Socializing also doesn't just help people in pain; it helps others understand people in pain. Canadian neuroscientist and pain expert Professor Jeffrey Mogil has dedicated much of his research to understanding the social mores of mouse society. In a well-known 2006 study, he showed that mice demonstrate more pain when they are in a cage with another mouse in pain, compared with when they are alone.[15] Importantly, this was only when the mouse was familiar to them. Subsequent studies have shown that humans demonstrate this pain contagion; similarly, their empathy and understanding of others' pain is much weaker when they see strangers in pain, as we saw in Chapter 7. Mogil's team has demonstrated that we don't empathize as well with strangers because we find them stressful; we produce more stress hormones (called glucocorticoids) when we are exposed to strangers, and this dampens down our ability to empathize. How to overcome this? In 2015, Mogil and his team found an effective treatment: the video game *Rock Band*. In their study, they confirmed that empathy for someone else's physical pain is weaker if the person in pain is a stranger, but this is overcome if they pharmacologically block the viewer's stress hormones. Even more importantly, if the strangers played *Rock Band* together – in which they strum and drum on the controllers to simulate being The Beatles – for just fifteen minutes, the two once-strangers now exhibited pain-empathy towards each other. The aim of the researchers clearly isn't to get us all to go out and buy *Rock Band*. It's about visiting someone living with persistent pain and having a chat over a cup of tea. It's going for a walk in the countryside, knitting or playing cards with them. Supportive, empathetic physical touch is also a powerful pain reliever.[16] Never underestimate the power of a good hug. Meaningful social interactions benefit people in pain but also give others understanding and insight into that pain. For both parties, stress is reduced, and this can only be a good thing for blood pressure, immune-system functioning, mental health and a love and openness towards others, particularly strangers.

Social interactions are often better than medications. These interventions are simple, cheap and benefit everyone involved. The way medical professionals treat people in chronic pain must change, and, thankfully, governments and medical bodies are taking note. 'Social prescribing' – where health professionals can refer people to support in the community – is beginning to gain traction in a number of countries.[17] But helping the lonely and isolated is down to each of us. The reality of isolation has now been experienced by us all, due to the Covid-19 pandemic, and I hope that the relatively short but shared time in isolation will have driven everyone to look out for the chronically isolated and lonely members of society.

Where there is social injustice, there is unnecessary pain. A 2016 study that analysed millions of emergency-department visits across the US found that black men and women admitted as patients are half as likely to receive pain medications as white patients.[18] If they are lucky enough to be given analgesia, they also tend to receive lower doses than their white counterparts. Black children coming into hospital with appendicitis are considerably less likely to be given any pain relief for moderate pain, or opioids for severe pain, than white children.[19] What is equally disturbing, however, is the extent of beliefs and biases that lead to these disparities. Many of these have been uncovered in recent years by an inventive team of social psychologists at the University of Virginia. In one of their studies, over six hundred medical staff from the National Collegiate Athletic Association (NCAA) were given a case study of a student athlete who has torn their anterior cruciate ligament, and were randomly assigned to the case for either a black or a white athlete.[20] The staff were first asked a number of questions about the case, which included rating the athlete's pain on a scale of one to four. They were then asked questions from the Symbolic Racism 2000 Scale, which measures racial attitudes by asking how much one agrees with prejudiced statements such as 'It's really a matter of some people not trying hard enough; if blacks would only try harder they could be just as well off as whites.' The more strongly one agrees with the racist statements listed, the lower one's score. The medical staff rated the black athlete's pain as less than the white athlete's. Surprisingly, there was no link between a participant's racial attitudes and their perception of the black athlete's pain. 'Even participants who have very positive racial attitudes show this bias,' argued Sophie Trawalter, one of the study's authors. The bias was focused around the athlete's initial pain, and not their recovery from pain, so it appears that the staff assumed that black people feel less pain, not that they cope with it better. The study cannot go as far as finding out *why* this systemic bias exists, but emphatically shows that it does.

Another study by this team involved over two hundred white medical students reading two case studies, of a white patient and a black patient each with a painful medical condition.[21] They were then asked to report the degree to which they agreed with scientifically incorrect statements about biological differences between black and white people, such as 'Blacks' nerve endings are less sensitive than whites' and 'Their skin is thicker than whites'.' The study found that medical students who endorsed these false beliefs tended to believe that the black patients felt less pain. Interestingly – and even more worryingly – those who didn't hold any of the false beliefs actually thought that the black patient was in more pain, but they didn't recommend they have more pain relief. Shockingly, half of the medical students agreed with one of the false beliefs. If this isn't a call to change health beliefs and acknowledge biases about race, then I don't know what is. And it's not just about the need for a paradigm shift in what's being taught; it's about who isn't being taught. There were fewer black men in

US medical schools in 2014 than in 1978.[22] It's not that medical schools are segregated and openly hostile. Instead, bias and racism permeate all structures of society to bring this about – from educational disparities, steep medical-school costs to a lack of role models, to name a few. When the medical profession of a country does not reflect its population, dangerous assumptions, biases and power imbalances can fester, causing real, horrible and unnecessary pain.

In 2019, I shared a panel with writer and campaigner Caroline Criado Perez, whose book *Invisible Women: Exposing Data Bias in a World Designed for Men* relentlessly exposes – with study after study – how women's pain has been, and still is, ignored. She opened my eyes to inequalities in the medical world that had been right under my nose but completely off my radar. There are still deep, male-centric cultural attitudes in medicine: 'instead of believing women when they say they're in pain, we tend to label them as mad', she writes in *Invisible Women*. While women are no longer locked up in asylums and given hysterectomies and even lobotomies for 'hysteria' (yes, this happened throughout much of the twentieth century), the fact that women are less likely than men to be given pain relief but more likely to be given sedatives and antidepressants shows that these attitudes are alive today.[23]

A 2008 study found that women admitted to A&E with abdominal pain are significantly less likely to be given opioid pain relief (such as morphine) than men, and that they wait longer to receive it once prescribed.[24] This is the case despite evidence showing that – among many of the underexplored sex differences in pain perception – women tend to require a higher dose of morphine than men to reach the same level of pain relief.[25] During my first two years as a junior doctor, I spent time in general surgery (essentially surgery anywhere along the gastrointestinal tract), which involved assessing hundreds of women in A&E, trying to work out whether their abdominal pain was caused by a 'general surgery' issue or 'gynaecological' disease, and whether they needed to be admitted to our wards or our theatres. Looking back, the findings of these studies sadly do reflect real life: whether subtly or starkly, women's pain is often dismissed. This was particularly true when a woman was suffering with pelvic pain that couldn't be easily seen on a scan or immediately cut out with surgery. The fact that women are less likely to be given pain relief than men is particularly cruel, given that women have (on average) lower pain tolerances, report higher pain severity and experience pain for longer than men.[26]

This is before we even begin to look at painful conditions that mainly affect women. Over the course of their lifetime, 90 per cent of women experience premenstrual syndrome (PMS), a poorly understood and under-researched collection of symptoms that include headaches, breast tenderness and abdominal pain. By comparison, 19 per cent of men experience erectile dysfunction at some point in their lives, but there are over five times more studies on erectile dysfunction than on PMS.[27] Endometriosis is another reasonably common but poorly understood condition, in which tissue from the uterus grows elsewhere in the body. It is often brutally painful and it is not unusual for sufferers to spend years bouncing between different doctors, completing a sticker book of incorrect diagnoses along the way, before being diagnosed with endometriosis; in fact it takes an average of eight years to reach a diagnosis of endometriosis in the UK, and ten years in the US.[28]

Delays in diagnosis are often life-ruining, but they can also be life-threatening: a 2018 Swedish study found that women experiencing a heart attack wait an hour longer than men from the first pain to arrival at hospital, and then wait a further twenty minutes longer than men to be seen.[29] There are many reasons for this, and these are probably socially entrenched, ranging from friends, family and even medical staff assuming that the pain isn't serious, to the female patients themselves 'not wanting to trouble anyone'.[30] In 2018, a twenty-two-year-old woman in Strasbourg, France, called the emergency services with terrible abdominal pain. 'I'm going to die!' she told the operator. The response came back: 'You'll definitely die one day, like everyone else.' When she was eventually brought in to hospital, five hours later, she suffered multi-organ failure and died.[31] The experiences and injustices of the pain of half the population are now beginning to be heard and recognized by government and health departments – better late than never. In 2017, England's National Institute for Health and Care Excellence (NICE) released its first-ever guidance to doctors on endometriosis, highlighting the need for doctors to listen closely to women. Lone Hummelshoj, chief executive of the World Endometriosis Society, summarizes: 'I suspect the overall message to physicians is to use their ears more than anything else and listen to what women are telling them.'[32]

'Listen to me' should be a mantra for any group that is ignored or oppressed by society. I have homed in on race and sex; if I were to do justice to all the groups and individuals whose pain is worsened by social injustice, it would fill a library, let alone the rest of this book. One crucial element that lies at the heart of social pain is the fact that perceived injustice worsens pain. If you think – or, in most cases, know – that you are being treated unfairly, your pain will be worse. The scenario that most commonly comes to mind is the car-crash victim whose persistent pain is (often justly) blamed on the reckless driver who caused their injury. The sense of injustice, however, does not have to be directly aimed at the initial cause of the pain; think of the person with the broken arm who focuses on their unfairly long wait in A&E or the time someone might spend grappling with a dismissive insurance company following an accident. But, even more importantly, the injustice doesn't have to be related to the pain at all.[33] One study in 2016 asked 114 healthy subjects to place their hands in ice-cold water (kept at a steady temperature) and rate the level of pain they perceived. Some of the participants were then asked to recall a time when they were treated unfairly before placing their hands in the water. Those who were thinking about injustice experienced more pain.[34]

It doesn't take a great stretch of the imagination to see how perceived injustice is petrol for pain's fire; a sense of injustice leads to rumination, anger, anxiety and stress, all of which propagate cycles of negative thought and behaviour that worsen persistent pain. Perhaps unsurprisingly, people who have strong 'just-world beliefs' – that the world is fundamentally just and that everybody eventually gets what they deserve – suffer more pain at the hands of perceived injustice.[35] And here we come to a paradox: we want to fight and ultimately eliminate injustice, and we want to reduce a sufferer's persistent pain, but the thoughts, actions and emotions required to fight injustice often make persistent pain worse. Evidence is emerging that often the best ways of calming pain related to injustice are through therapy that enables psychological flexibility and acceptance.[36] Acceptance, importantly, is not giving up and giving in. It is enabling, as someone is helped to understand their situation and to see what, as an individual, they can and can't change. It is difficult and there are no easy answers. But this is not brushing inequalities under the carpet: injustice needs to end. And, if anything, this should encourage others – those not living with persistent pain – to actively advocate for those who are. We should all support and give a voice to those who have been wronged and hurt. They should never have to go it alone.

'Did you know that, up until the late 1980s, it was widely believed that babies didn't feel pain . . . Babies were having operations without pain relief. What's worse is that they were being given neuromuscular blocks but no analgesia, so the babies would be paralysed, conscious and still feeling pain.'

I had just met Deniz Gursul, who, at the time, had just started a PhD in 'baby pain'. I had almost completed medical school at that point, yet this conversation was the first time I had ever considered pain in infants. I later interviewed Dr Gursul at the University of Oxford, after she had spent several years using brain-imaging techniques to investigate infant pain. I was curious as to why the medical establishment had assumed that infants didn't feel pain. Surely every parent – every human – knows that babies can hurt?

Dr Gursul told me that the foundations for infant pain denial had been laid in the early twentieth century. Researchers would assess infants' pain levels from their response to pinpricks and electric shocks.[37] 'When I look at literature from these times,' she told me, 'I can't believe it . . . the studies are ridiculous. They'd get a pin, poke a baby's foot and report: "No overt response."'

The researchers observed that infants seemed to respond slower to pinpricks than adults did. But even when infants responded violently, this was put down to primitive reflexes, not pain. The observations were coloured with the assumption that babies' nervous systems were not fully developed, the knowledge that we don't have explicit memories of our very early childhood and an understandable fear of overdosing infants with anaesthesia. This perfect storm of questionable assumptions and anaesthetic caution meant that the supposed insensitivity of infants became established medical canon and was engraved in the textbooks during the course of the twentieth century.

It took an angry and heroic mother to shift the paradigm. It was 1985, and Jill Lawson had just given birth to her son Jeffrey at the Children's National Medical Center in Washington, DC. Jeffrey had been born prematurely and had to undergo open-heart surgery. It wasn't until after the operation that Jill found out her son had only been given a muscle relaxant and received no pain relief. She was even more shocked to find that this was standard practice. Jill's letters and pleas were ignored, and her fears played down by medical staff. That was until the *Washington Post* published her story the following year.[38] Jill's advocacy lit the flame for some desperately needed research. A year later, in 1987, a team at Oxford University conducted a game-changing study. The research assessed preterm babies undergoing open-heart surgery, comparing those given fentanyl (an opioid pain reliever) with those given no pain relief at all.[39] The babies given pain relief had much better outcomes and fewer complications, and blood tests showed that their bodies experienced less stress in response to the surgery. Thanks to impassioned parents and inventive scientists, we have come a long way over the past three decades. The International Association for the Study of Pain (IASP) now states: 'Verbal description is only one of several behaviours to express pain; inability to communicate does not negate the possibility that a human experiences pain.'[40]

But if babies can't tell us that they are in pain, I asked Dr Gursul, how exactly do we know they're in pain? 'Technically, we can't fully know,' she told me. 'Pain is a perception, and it's hard to tell what someone perceives if they can't tell us. But we can approximate a level of an infant's pain using "surrogate measures". Some of these are quite crude, such as measuring the baby's cry – its length, pitch and duration. Part of my PhD work involved observing babies for thirty-second periods and recording when they made specific facial movements, like brow bulges and eye squeezes.'

Baby-pain research certainly isn't for the faint-hearted. Dr Gursul pointed out the fact that babies cry for almost any reason, and this is why new advances in brain-imaging technology have been so important for measuring infant pain. Two non-invasive methods that she has used are functional magnetic resonance imaging (fMRI), which uses changes in blood oxygenation in the brain to see which areas are active, and electroencephalography (EEG), which uses electrodes on the scalp to measure how electrical activity in the brain changes with time. It's not quite mind-reading, but neuroimaging is a huge step forward in understanding the feelings of the voiceless. Much of this work has been carried out in the Oxford lab of Dr Gursul's supervisor, Professor

Rebeccah Slater. In 2015, an fMRI study by Slater's team found that eighteen of the twenty brain regions that are commonly active in adults during pain are also active when infants are exposed to a noxious stimulus (don't worry – researchers use completely harmless retractable probes that don't penetrate the skin), which suggests that infants have the potential to experience pain in a similar way to adults.[41] In fact, it was apparent in the research that infants' brains were even more responsive than adults' in response to stimuli. This means that for most of the twentieth century the medical establishment had assumed that babies couldn't feel pain, whereas in fact they might be even more sensitive than adults. In another study, the same team in Oxford also used EEG – electrodes that measure brainwaves – and found that a noxious stimulus produces pain-specific electrical activity in an infant's brain and that this is strongly related to the strength of the stimulus.[42]

It is evident that babies experience pain. While pain is clearly distressing in the short term, there is also evidence that early-life pain has lifelong effects. Repeated procedural pain as a baby is associated with abnormalities in pain processing as well as behavioural disorders and reduced cognition (such as intelligence and linguistic abilities) in later life.[43] It is important to note that 'associated with' does not necessarily mean a causal relationship, but it is reasonable to infer that repeated childhood pain may affect the development of the pain system and brain as a whole. It's also tempting to speculate that extensive early-life pain could lead to chronic pain conditions much later down the line, but the complexities of long-term data-gathering and accessing detailed neonatal records mean we don't yet have any data to show that relationship. Regardless, we have a duty to ease the pain of this extremely vulnerable group. And there's a lot of work to be done. A 2014 review found that babies in neonatal intensive care units experience about a dozen painful procedures a day – from tubes inserted up noses and down windpipes, to heel-prick blood tests – yet the majority of babies do not receive any kind of pain medication.[44]

Professor Slater's team are helping to pioneer the use of neuro-imaging to see which methods and medicines can actually relieve pain in infants. They found that applying anaesthetic cream to a baby's skin before a procedure reduces pain-related electrical activity detected by EEG.[45] But not every analgesic has been successful. The group tried to assess whether giving morphine to premature infants before they had painful procedures, such as blood tests and eye examinations, relieved pain, but unfortunately not only did morphine not appear to reduce any of the pain measures compared with a placebo, but also the trial had to be stopped early due to morphine-related side effects.[46] It's important to note that, as the study was small, we can't draw too many conclusions from it. But the team did come across a pain reliever that is readily available, inexpensive, low-tech and side-effect free: touch. We all know the comforting, pleasant sensation of a reassuring pat or stroke. The science behind stroking is well established: if you want to achieve the perfect skin-on-skin caress, lightly stroke someone's skin (with their permission, please), travelling around three centimetres per second.[47] This activates 'C-tactile fibres' in the skin, which send the brain signals associated with pleasant social touch.[48] In adults, this stroking successfully reduces the perceived intensity of short-term pain.[49] Dr Gursul wanted to see if something similar happened in infants. Just before babies were subjected to a noxious stimulus, they were either stroked with a brush moving at the optimum speed of three centimetres per second, stroked too quickly – at thirty centimetres per second – or not stroked at all.[50] Dr Gursul found that stroking at the optimal speed did reduce babies' pain-related brain activity, whereas the other controls didn't. This gives us confidence that gentle, reassuring touch really does relieve pain in infants. It feeds an infant's brain with positive information and clearly conveys an impression of safety. It adds to a large body of evidence that touch – from skin-to-skin contact with a premature baby (known as 'kangaroo care') to massage at any age – conveys real health benefits. Our understanding of infant pain is itself in its infancy, but it shows great promise for relieving the unnecessary suffering of every new human setting out on the journey of life.

Pain is social. Pain is almost always worsened in those hurt by society: the lonely, the marginalized and the voiceless. It's remarkable, but perhaps not surprising, that social structures worsen pain through the same means by which torturers manipulate environment and psychology. Isolation, humiliation, intimidation, oppression and unfairness may seem like abstract concepts, but they all worsen the physical and emotional experience of pain. This makes sense in light of the fact that pain is our protector: it is soothed by safety and inflamed by threat. A modern understanding of pain should encourage us to look out for the vulnerable and look after the oppressed. Pain should drive us to love.

9

Belief as Relief

Faith and frameworks

If a man says he is not afraid of dying, he is either lying or he is a Gurkha.

FIELD MARSHAL SAM MANEKSHAW, CHIEF OF THE ARMY STAFF IN THE INDIAN ARMY, 1969–1973

A CLOSE FRIEND OF mine is Mancunian and one of his more endearing terms for me is 'Soft Southerner'. The British stereotype – that your pain tolerance and general toughness is directly related to your latitude – is an old one. Indeed, the idea of cultural differences in pain perception and tolerance is as old as humanity itself. We all have impressions and opinions, many of which are fed by the hearsay and stereotypes created by our own culture. I grew up voraciously reading any history book I could find, and particularly ones about ancient, alien peoples who had seemed to master pain and fear, from ferocious Amazon women of Greek legend to battle-hardened Vikings. One group of people for whom I've always had a romantic reverence are the Gurkhas. These are Nepalis (or ethnic Nepalis from India) who serve in a number of militaries around the world, including the British Army. These soldiers have a legendary reputation for bravery. They are recruited in the mountains of Nepal, where as many as 25,000 tough young men apply for around two hundred places annually. The selection culminates in the gruelling Doko Race, where contenders run five miles up the side of a mountain carrying a twenty-five-kilogram basket strapped only to their heads.

Given that Gurkhas and British officers serve and fight in the same battalions, I've always been curious as to whether the former have higher pain thresholds and tolerances ('pain threshold' being the minimum intensity at which a stimulus feels painful, and 'pain tolerance' the maximum pain someone can bear). My research revealed no studies; the closest I came was with a relatively well-known 1980 study that found that Nepalese mountaineering porters had much higher pain thresholds than their European counterparts.[1] In the absence of existing data, I decided to consult someone who has been intimately involved in Gurkha selection his whole life. Colonel James Robinson, CBE, was born in Nepal (the son of a Gurkha officer), joined the Royal Gurkha Rifles and was the head of the Brigade of Gurkhas between 2012 and 2019. He said: 'I don't see any noticeable differences in pain thresholds between Nepali and British soldiers today. But I think there were in the past. When I first went to Nepal to recruit soldiers, in the 1980s, many of the villages we'd travel to weren't even accessible by road. The people were mostly subsistence farmers and the young men were tough. They had a greater acceptance of pain than the Westerners, who at the time were definitely 'softer'. But the gap has been closing as Nepal has been Westernized. In the 1990s, young Nepalis would go to India and the West to train as doctors, bringing back a largely pill-based system. Roads have been built to the villages and Western comforts and medicines are now much more available to these young men.'

Colonel Robinson's observations reflect the broader fact that we live in a shrinking world, and what were once distinctive national and cultural borders are becoming increasingly blurred. In many ways this is a good thing; it should encourage us to treat each person as an individual, not based on assumptions about their background. Colonel Robinson made another observation while leading the training of Gurkhas: when Nepali soldiers arrived in Britain for training, they tended to go to the regiment's medical centre with training injuries only when these became very serious. This could be due to their high pain thresholds or, more likely, that they didn't want to be seen as injured, for fear of having to leave the regiment and go home. An individual's perception of pain is linked to the way in which they communicate that pain to others, and this varies widely across cultures. An extreme example can be found in the Bariba people in West Africa, who take stoicism to another level. The Bariba have almost no vocabulary for pain, and instead indicate their virtue through not expressing pain at all. Women are expected to endure childbirth silently and men to take the wounds of war without complaint, lest they bring shame on their family.[2] Honour–shame culture is not the preserve of the Bariba, however. Only over the past century has British society moved from a culture of 'stiff upper lip' to 'express yourself'. While the stigma of disease and pain still remains, it is now much more socially acceptable to share and express your suffering.

The comparison of pain thresholds of different cultural or ethnic groups has a long scientific history. In 1965, the Harvard psychiatrists Richard Sternbach and Bernard Tursky subjected American 'housewives' of varying ethnicities to electric shocks. They found that there was no difference in sensation threshold – the level of shock needed to report a sensation – but found significant differences in pain tolerance. For example, Italians tended to have lower pain tolerances to electrical stimulation than 'Jews' and 'Yankees'.[3] This may give insight into varying cultural attitudes towards pain at the time: one commonly suggested explanation was that Italians are more expressive and open to communicating their pain, whereas Americans of northern European descent want to appear more reserved and stoical. But these studies show patterns, not causation, and lend themselves to sweeping generalizations and the feeding of stereotypes. Cultural and ethnic attitudes to pain are diverse not only between, but within, groups, and they also change across time. Colonel Robinson's experience with Gurkha soldiers suggests that the attitudes Sternbach and Tursky saw in these groups in the 1960s may be very different today.

A 2017 study that analysed much of this literature, and statistically combined the findings of the more robust studies, found that most ethnic minorities in the USA and Europe tend to be more sensitive to pain than white individuals.[4] But, interestingly, when these groups are tested in a country in which their ethnicity is the majority, these ethnic-group differences are not apparent. In one study, Indians tested in India had higher pain thresholds than individuals of Indian ancestry (second generation or higher) in the USA.[5] It seems that being part of a minority – and all the health and social injustice this entails – is the main mediator of pain. This has been strongly supported by a large Swedish study, published in 2020, that analysed questionnaires completed by over 15,000 people from a random population sample.[6] It found that immigrants (regardless of ethnicity) had much higher levels of chronic pain, widespread pain and severe pain than Swedish-born participants. In this study, it appears that the main factors that link immigration status to chronic pain are depression and anxiety. These findings are supported by a 2019 American laboratory study that found that anxiety, depression and stress are significant causes of heightened pain sensitivity among ethnic minorities.[7] This makes sense when we come back to the essence of pain: a feeling trying to protect us. Minorities and migrants are often in a position of vulnerability and individuals in these groups often live with an understandably heightened sense of threat. This leads to fear, stress and depression – a perfect storm for pain.

Cultural and ethnic differences in pain perception and communication teach us a few crucial things. First, we need to recognize that these differences are hugely complex and fluid, and we should never make generalizations about a group and assumptions about an individual because of their background. A nursing textbook published in 2014 got this badly wrong.[8] Here are two (of many) statements made in its section on cultural differences in pain beliefs: 'Jews may be vocal and demanding of assistance' and 'Blacks ... believe suffering and pain are inevitable.' Regardless of the textbook's intentions, this is racist, wrong and – if it influences the outlook of just one nurse – dangerous. After an understandable uproar, the publisher removed the material and apologized. If we are comparing ethnic or cultural groups, it needs to be done in a way that develops respect and cross-cultural understanding, and ideally helps manage pain for everyone in our diverse society. Cultural attitudes towards pain vary wildly. For instance, my own 'cultural attitudes' towards pain are probably very different from my brother's, I being a doctor and he being an army officer. More than anything, we need to see people as unique individuals. Second, we have to also bear in mind that, while we are all individual, it is clear that ethnic and cultural minorities in a country are particularly vulnerable to chronic pain and pain injustice. And as a society, we need to pursue anything that reduces the alienation of minorities.

Variations in pain experience, tolerance and expression are of course influenced by culture, but it ultimately comes down to what pain *means* to us. And what meaning is stronger than our beliefs about the meaning of life, the universe and everything? In 2008 Professor Irene Tracey – Oxford University's 'Queen of Pain' and one of the greatest contributors to pain research in the last few decades – gathered an unusual pairing of subjects for an experiment: practising Catholics and avowed atheists.[9] Each participant was exposed to a series of electric shocks while lying in an fMRI scanner. Initially, they simply lay in the scanner and received shocks. Later in the experiment, thirty seconds before each shock was delivered, the participants were shown one of two Italian paintings – either Sassoferrato's *The Virgin in Prayer* (of the Virgin Mary) or Leonardo da Vinci's *Lady with an Ermine*. The paintings stayed in full view when the shock was administered. The results showed that while both religious and irreligious participants had very similar baseline perceived pain-intensities when shocked in the absence of any images, Catholics reported significantly lower pain-intensities when viewing the image of *The Virgin in Prayer*. The fMRI results backed this up: when Catholics were given electric shocks while looking at the image of Mary, an area of the right ventrolateral prefrontal cortex lit up. This is a region associated with inhibiting the strength of danger signals travelling to the brain from the body. And while the religious element of pain is certainly not an over-researched field, a 2019 review found that a number of other studies suggest that religiosity and spirituality help with handling pain, and can even reduce pain-intensity.[10] It seems that belief can bring relief.

While few converts will be made by the fact that religious belief has some effect on soothing pain – the avowed atheist Karl Marx called religion 'the opium of the masses'[11] – this is certainly something worth exploring. At my medical school, there were a number of 'token' lectures or modules – elements of teaching we were told were important for patient care but ultimately would never come up in exams; they were therefore sadly never given access to the precious space in our memories. One of these subjects was the 'spiritual dimension of health': patients' deep-seated beliefs and how these influence their healthcare. Doctors dismiss the spiritual dimension of disease at their peril. It's particularly important for two key reasons. First, the majority of the world's population is religious, and to ignore an individual's deepest beliefs is arrogant and wrong. Second – specifically for pain – if there are elements of religious belief that soothe pain, surely these can be of help for the secular, post-Christian West. Humans have always been grappling with the problem of pain, and the wisdom of the Ancients may well have a few lessons for us.

Paul Brand was a pioneering leprosy physician. On graduating from University College Hospital Medical School during the Second World War, he was put to work as a casualty surgeon in the

London Blitz. Straight after the war, Dr Brand took his skills to Tamil Nadu, India, where he would spend the next twenty years. Alongside introducing tendon-transplant techniques that enabled leprosy sufferers to regain the use of their hands and feet, he conclusively showed that the tissue damage and deformities of leprosy were not directly caused by *M. Leprae* bacteria, but that instead these microbes damage the danger-detecting nerves in the skin. The resulting lack of pain ultimately leads to tissue damage. His experience of understanding the importance of pain for life led him to write his best-known book, *Pain: The Gift Nobody Wants.*[12] He also noticed that the deeply religious people of this area of India – whether Hindu, Christian or Muslim – seemed more equipped to live with pain than those in the Western culture he had left. Brand put it down to a number of things: acceptance, gratitude, prayer or meditation, and strong family networks. But in 1966, after two decades of working in India, his move to the United States starkly revealed this cultural chasm: 'I encountered a society that seeks to avoid pain at all costs. Patients lived at a greater comfort level than any I had previously treated, but they seemed far less equipped to handle suffering and far more traumatized by it.' His theory, fleshed out in the books and lectures of the later part of his life, is this: the West's pursuit of happiness and pleasure as the main 'good', and our partial success at relieving pain with science and medicine, has paradoxically made us worse at living with pain.

While Dr Brand's theory is just that – a theory – I think elements of it are true. Pain has become an enemy: an unseen assailant that we fight with pain*killers* and by recruiting doctors and surgeons to the battlefield of our body. In a purely secular society, pain is, at best, a random interruption of our pursuit of a life of happiness, pleasure and freedom. Pain doesn't have any meaningful part in our story. At worst, pain is a blind, purposeless force of life-ruining evil. While minimizing physical pain, particularly persistent pain, is a noble aim, it is clear that viewing pain as an enemy to be avoided or killed is profoundly counterproductive.

While the great world religions clearly differ from each other in numerous ways, a commonality to all these beliefs is that pain plays a part in achieving our purpose, whatever that may be. I think this can be broken down into two elements – two elements so often lacking in the West: acceptance and hope. Religions do not beat about the bush when it comes to the reality of suffering and the fact that pain is an integral part of human existence. Most religious texts are infused with pain, and many are solely dedicated to explaining it. A powerful example of acceptance is found in Islam, in which submission to Allah's will is central. I asked a Muslim friend and colleague, Dr Ithsham Iqbal, to summarize how he views pain as a Muslim: 'Islam teaches us that Allah is merciful and gives Muslims trials such as pain as a way for us to actually get closer to Him. It is a source of contentment that Allah has given us these hardships because He wants us to become better Muslims. There is also reward for going through trials. Islam teaches us that by placing our faith in Allah when we go through difficulties, we actually increase in our good deeds. The reward ultimately brings peace in knowing that pain is given as a blessing rather than as a punishment.'

In Islam, the acceptance of the reality of pain is clearly a necessary step for self-growth and mastery – and endurance is key to this. A scripture frequently shared by believers on social media encourages fellow Muslims to view pain as having a purpose: 'When Allah desires good for someone, He tries him with hardships.'[13] This in no way means that Muslims are expected to passively give in to pain; alongside this understanding that pain is inevitable, there are Muslim doctors and researchers who pioneered the fields of pain relief and anaesthesia as far back as the tenth century.[14] That said, each member of a religion or religious culture is an individual, with their own experiences and beliefs, and no one should ever be stereotyped. What we're potentially learning from, however, are religious approaches to pain that could serve us all.

Two other ancient beliefs – both born around the same time, but one formed in the schools of ancient Athens and the other in the towns of the Ganges plain – converge on their understanding of pain acceptance. The Ancient Greek Stoics believed in accepting the sensations of the moment for what they are, and not attaching value judgements to them, whether good or bad. This idea of acceptance and detachment is also central to Buddhism. A wonderfully insightful teaching on pain, attributed to the Buddha himself, is termed 'the second arrow': 'When touched with a feeling of pain, the ordinary uninstructed person sorrows, grieves and laments, beats his breast, becomes distraught. So he feels two pains, physical and mental. Just as if they were to shoot a man with an arrow and, right afterward, were to shoot him with another one, so that he would feel the pains of two arrows.'[15]

This is early, astute knowledge that pain is not simply sensory, but also emotional and cognitive. We can't control all of our circumstances, but we can learn to control our reaction to them. This thinking has powerfully influenced a number of 'modern' ways of treating and managing persistent pain: mindfulness, cognitive behavioural therapy, acceptance and commitment therapy, and hypnotherapy, to name just a few. While acceptance is certainly not a cure for most people living with persistent pain, it is a necessary first step. It seems paradoxical that accepting pain rather than fighting it makes it more bearable, but it makes complete sense in light of what pain is: a system that is part of us, trying its best to protect us.

Alongside accepting the reality of the present situation, many religions provide hope for the future. Dr Iqbal told me that, through suffering pain on this earth, Muslims can become closer to Allah and achieve future relief in another life. In karmic religions, such as Hinduism and Buddhism, pain drives believers to do good and live moral lives, which will ultimately lead to less pain in the next, reincarnated life.

Pain is also utterly central to Christianity. Indeed, the primary Christian symbol – the symbol that has formed the shape of cathedrals, appeared on national flags and even adorned spiced buns – is an implement of physical torture. The Roman cross is designed to kill with as much pain, suffering and humiliation as possible. Christians believe that Jesus (fully God yet fully man) was tortured and died on a cross to pay the penalty God justly demands for human wrongdoing – the punishment we deserve. Christians take comfort in the fact that, although they experience pain in this life, they have a God who understands and empathizes with their pain because he himself had nails driven through his hands and feet. Peter, one of Jesus' followers, tells early Christians – most of whom experienced persecution and torture – that when they experience physical pain, they 'share in Christ's sufferings'.[16] As well as taking comfort in the present, Christians believe that Jesus' death and subsequent resurrection ultimately defeated death and pain. They have a certain hope for a future in which – outlined in one of the very last sentences in the Bible – 'There will be no death or mourning or crying or pain.'[17]

I asked Tim Keller, a Christian theologian and writer, and pastor of Redeemer Presbyterian Church in New York City, what Christianity has to offer the person in pain. 'Christianity offers a lot of unique resources to persons in pain. First, it is the only major faith or world-view based on the understanding that God himself entered the world and saved us through his own suffering. Jesus, the Son of God, did not triumph over evil in spite of his weakness and suffering, but through it. And that gives Christians a unique framework for understanding suffering as not merely a waste, but as a means to greater wisdom, beauty and the overcoming of evil. Second, Christianity is the only major faith or world-view offering a vision of a renewed material world. Christian salvation does not merely redeem souls for a trouble-free spiritual afterlife. It redeems bodies for a new physical creation, so we have not just a consolation for the world we wanted but never had. We *get* the world we wanted but never had – free from stain or blemish, from suffering, decay, and death.'

Tim is no stranger to pain himself, and he has certainly had to put his faith into practice. At the age of fifty-one, he was diagnosed with thyroid cancer and underwent radiotherapy and surgery. At the same time, his wife's Crohn's disease had become particularly acute; she required seven operations in one year. When I interviewed him in late 2020 he had been recently diagnosed with Stage Four pancreatic cancer. Full recoveries are rare.

'This has of course moved my wife and myself into the most extensive and intense time of reflection and prayer that we have ever had in our lives. We have re-visited all that we have learned from the Scriptures over the years about suffering and found it more than enough to support us and even to give us more poignant joy in the midst of tears than we've ever had.'

Whether we – or those we are caring for – are religious or not, we need to recognize the centrality of belief to pain, particularly persistent pain. We can also learn valuable lessons from belief – lessons that can help us live with pain and even soothe it. Acceptance and hope may seem like contradictory ideas, but holding on to both of these is key for living with and soothing pain. Acceptance is seeing the difficulties and life-changes for what they are, and knowing that the pain will not be going away any time soon. Hope is not simply wishful thinking; it is knowing that, despite the problems of the present, there is real evidence that persistent pain can get better. Reframing one's beliefs to have an ultimately positive outlook is easier said than done, but it really works. This is best learned in a supportive pain-education environment, but we can start this process of reframing by looking at the most important beliefs concerning pain. We can either believe that persistent pain means that our body is damaged, or we can believe the enormous evidence from modern pain science – that pain is our protector and most cases of persistent pain are caused by our overprotective pain system in the brain.

Let's take the case we've begun to build against low back pain as a practical example. Our spinal column is immensely strong, flexible and adaptable, and, like the rest of our body, it is very good at healing. But we in the West have accepted the belief that our backs are fragile, with discs on the verge of 'slipping' and nerves just waiting to be 'pinched', all held together by 'crumbling' bones. This view is the result of many factors. It's not helped by the use of outdated biomechanical models in medical and physiotherapy training. Also, there are plenty of people whose careers depend on backs being 'misaligned', but I won't go down that rabbit hole. The reality is this: the vast majority of cases of low back pain are not caused by permanent tissue damage, and there is a poor relationship between low back pain and evidence of tissue damage. People can have terrible pain yet nothing shows up on any scans; equally, plenty of healthy, pain-free individuals happen to have dodgy-looking scans while feeling no pain at all.[18] In fact, 37 per cent of pain-free twenty-year-olds and 96 per cent of pain-free eighty-year-olds have signs of 'disc degeneration' on scans.[19] These harmless changes that increase with age are as medically unimportant as wrinkles. Most cases of long-term low back pain, however, are due to an overprotective brain trying to protect a healthy spine. This doesn't mean that the pain isn't horrible or isn't real; just that pain has become wired in the brain. It's clear that believing that the pain equals damage actually worsens the pain, whereas treatments that provide knowledge, confidence and hope really can cure.

One treatment is cognitive functional therapy (CFT), which aims to change the sufferer's mindset about low back pain through education, gradual exposure to controlled movement and tasks, and a healthier lifestyle. CFT's ultimate goal is to reframe someone's belief, changing it from a cycle of fear and avoidance to one of evidence-based confidence. And changing the narrative works: in a 2013 randomized controlled trial, CFT – in which participants took part in sessions and workshops that helped reframe their pain beliefs and gave them confidence to move – was much more effective at relieving long-term low back pain than manual therapy and exercise.[20] These results were still seen after a three-year follow-up in 2019.[21]

Meaning is medicine, and hope really can heal. The idea that our pain-beliefs can completely change the course of persistent pain is profound, but it can easily be misinterpreted or misunderstood without the right tools to hand. To understand how we can turn the tide of persistent pain, we now need to really understand what it is.

10

The Silent Pandemic

The persistent pain crisis

Turning and turning in the widening gyre
The falcon cannot hear the falconer;
Things fall apart; the centre cannot hold;
Mere anarchy is loosed upon the world

W. B. YEATS, *THE SECOND COMING*

EVERYONE, IN THEIR heart of hearts, would love to have something named after them. Some would choose a newly discovered species of butterfly; others, a planet. There are those for whom a well-placed park bench would do. I, like many other slightly unhinged doctors, wouldn't mind an eponymous disease. So, please humour me for one short thought-experiment: imagine that a terrible illness, 'lymanitis', has swept through your country. It is a long-term, life-ruining disease that destroys one's daily functioning – from memory to mental health, from sleep to sex life. It's not infectious but its prevalence is growing year on year, and today it is estimated that between one third and one half of your country's population has it. Lymanitis is one of the top causes of short- and long-term sickness absence from work, and drains billions from your country's economy. It's been around for a while, but it was only officially recognized as a disease last year, and medical students are taught about this disease for an average of thirteen hours in their six-year medical degree.

If you've read the chapter title, you win no prizes for guessing that lymanitis is actually 'persistent pain'. And my made-up disease doesn't just reflect persistent pain; it is a direct substitute – each of the aforementioned facts is true for persistent pain in the United Kingdom.[1,2] But this is not just a UK or Western problem; as a ballpark figure (epidemiological studies on this complex disease vary considerably) in most countries across the world, from the USA to the developing world, roughly one in five people live with long-term pain.[3,4] And all across the world, numbers are rising. Not only are we facing a persistent-pain crisis but we are also poorly equipped to handle it. Persistent pain is complex, unique to each individual and rarely solved by a pill or surgical procedure. Living with and diminishing this pain is a slow and often hard path. Doctors, by their nature, want to find a measurable, visible and treatable cause for a disease, and long-term pain doesn't fit into this box. Henry McQuay, Professor of Anaesthetics at Oxford University, sums up the situation pithily: 'Chronic pain is common – but it isn't sexy.'[5] Pain is low on most governments' priority lists and doesn't see anywhere near the funding put into cancer or infectious diseases.

At this point we must remind ourselves of an important clarification. 'Persistent pain' and 'chronic pain' are the same thing. 'Chronic' is more commonly used in medical circles, but we are thankfully seeing a shift away from this confusing (and somewhat disheartening) word to 'persistent'. But before we can tame the dragon, we need to define persistent pain and understand what it is, what causes it and what worsens it. We can then find out how to diminish it. Knowledge is power.

Persistent pain is pain that persists or recurs over a long period of time. It is essentially that simple, but there is disagreement over what constitutes a long period of time. The International Classification of Diseases (ICD) defines it as three months, but it's generally agreed that persistent pain continues for longer than the expected time-frame for a specific injury to heal. Of central importance – and something that is under-appreciated – is the fact that in the vast majority of cases of persistent pain the initial injury (if there ever was one) has healed. Persistent pain is pain that has stopped being a symptom and has become the disease. It has taken until very recently for this to reach international medical consensus: 'chronic pain' first appeared as a disease in its own right in the eleventh revision of the ICD, in May 2019.[6]

The pain pandemic is large, global and growing. But before we get to the roots of what is driving it, we need to see its effects played out in another tragic health crisis, the proportions of which cannot be overstated. Between 2015 and 2018, the USA experienced its first sustained drop in life expectancy in a hundred years.[7] A century ago, the First World War and the 1918 influenza pandemic were the culprits. Now it is overdoses and suicides, lives ended prematurely by the 'opioid crisis'. Opioids, as we have seen, are substances that cause pain relief through binding to opioid receptors in the brain. Our body has its own internally made opioids (such as endorphins), but humans have been extracting opiates from the opium poppy since the dawn of civilization. More recently, we have created completely synthetic opioids, the well-known ones being oxycodone (sold as OxyContin and Percocet) and fentanyl. The crisis began with a perfect storm in 1990s America: powerful pharmaceutical giants lobbying lawmakers, making fraudulent claims that denied side effects and funding free education courses for doctors with little pain-management experience. Throw into the mix doctors and private-healthcare companies receiving incentives for prescribing, direct-to-consumer advertising and a culture that has significant belief in pills and prescriptions, and it didn't take long before opioids were the most commonly prescribed drug class in the US and overdose deaths from prescription opioids started to mount. Over a couple of decades, opioid-overdose deaths tripled, and by 2018, for the first time in history, Americans had become more likely to die from accidental opioid overdose than from a car accident.[8] Crucially, the majority of these deaths are associated with a diagnosis of chronic pain.[9]

Opioids are fantastic drugs for short-term pain, and I've seen their magic countless times in A&E and on surgical recovery wards, but they are much less effective for persistent pain. For long-term musculoskeletal pain (such as low back pain and osteoarthritis) they are no better than regular pain relievers such as paracetamol.[10] Professor Clifford Woolf – a Boston-based giant among pain scientists, who discovered the main mechanism for how short-term pain becomes long-term – summarizes the efficacy of opioids in long-term pain: 'The data is pretty overwhelming that the level of pain relief from chronic non-cancer pain is very low indeed.'[11] Over weeks, and even days, patients develop tolerance to opioids, in which the drug becomes less effective as the body's opioid receptors become desensitized, with larger doses needed to provide the same level of pain relief.[12] People on long-term opioids also often develop dependence – a need to have the drug to maintain normal function – and exhibit withdrawal symptoms – vomiting, diarrhoea, insomnia and sweating, to name a few – if taken off it quickly. Aside from the risk of accidental overdose, opioids also come with a Pandora's box of side effects. To make things worse, over a long period of time opioids can, paradoxically, increase sensitivity to pain, something known as opioid hyperalgesia.[13]

We shouldn't, however, throw the baby out with the bathwater, and there are obvious caveats. For example, opioids are useful (and often indispensable) for types of cancer-related pain. But while some people benefit from long-term opioids, most people don't. Doctors should rightly be cautious and controlled in their opioid-prescribing. In the UK, the Royal College of Anaesthetists' Faculty of Pain Medicine recognizes that opioids are rarely effective in the long term and now provides helpful advice for patient assessments, the conducting of initial trial courses of opioids (if necessary) to check whether they are effective, and for tapering and stopping them if they're not working.[14] We should also, of course, invest in new pain-relieving medications and those that enable opioid use without side effects or dependence. But there is a better way: we can live with, reduce and sometimes eliminate long-term pain without any reliance on opioids.

The pain pandemic and its contribution to the opioid crisis is a societal catastrophe, and it's growing year on year. Understanding what is causing this growth can give us clues as to how pain thrives and how we can soothe it, both as individuals and as a society. The modern world has a lot going for it, but it is fertile soil for pain. It's a world of social isolation, fast food, social media and sedentary lifestyles. It's a world of uncertainty and inequality; of risk aversion and fear. In short, it's a world of stress.

Stress is good. Or, should I say, short-term stress is good. To illustrate this, I considered describing a terrifying encounter between one of our ancient ancestors and a wild animal on the African savannah. Then it dawned on me that I have an old friend for whom that scenario is lived experience. It's a story that is hard to believe, and one that has put me off going on walking safaris for life. It was the school summer holidays and lucky, fourteen-year-old Henry was on safari in South Africa with his family. 'We were meant to stay in the four-by-four and our guide was meant to drive us around the game reserve,' he told me. 'But for some reason, he got us to step out of the vehicle and go deeper into the bush on foot.' With the guide at the front, Henry and his family walked single file across the savannah in search of the big five. 'Then, suddenly, the guide stopped and told us to stay silent. He'd seen a black rhino mother and some of her young not far ahead.' What happened next is a blurry memory for Henry. He recalls the frantic screams of the guide: 'Get behind a tree, get behind a tree!' As he spun around to find cover, Henry caught the most fleeting glimpse of a black spectre – a vision of death. The sum of human terror was condensed into one ton of angry mother rhino, bursting out of the long grass and charging straight towards him. As though possessed by another being, every muscle in the body of the boy who 'never, ever played sport' engaged in swift, violent action, carrying him away from the rhino's path in a dreamlike sprint. But it wasn't enough. Henry vaguely remembers being thrown up – completely painlessly – into the air and landing in a heap on the African soil. As the rhino stampeded into the distance, and the dust physically and figuratively settled, the group found Henry lying on the ground, blood oozing down his right leg and onto the dirt. He had been impaled by the rhino's horn, which had entered the muscle of his right buttock and – revealed later by a scan in hospital – had passed straight through it and into his abdomen. Miraculously, the horn had missed major blood vessels and organs, and he went on to make a complete recovery. Henry remembers the pain starting only perhaps twenty minutes after the event, in the back of the four-by-four on the way to the medical centre. The area around the wound would remain painful and inflamed well after he had been stitched up, however.

Stress had saved Henry. Those lightning-fast reactions he never knew he possessed had moved him a life-saving inch away from the centre of the rhino's path. The 'fight-or-flight' response makes us momentarily superhuman and it has been integral to human survival for millennia. Statistics suggest that you probably haven't been chased by a protective mother rhinoceros, but you'll recall having some of these symptoms minutes before an important job interview or speaking in public. In the acute stress response, there were three main body systems working synergistically to protect Henry. They are crucial to understanding how stress contributes to pain. We'll dive into the science in a moment, but the main takeaway is this: like pain, short-term stress is protective, but long-term stress is detrimental to our general health and is a driving force behind persistent pain.

The first of the three protective systems is the nervous system: Henry's whole fight-or-flight response was triggered by a largely unconscious recognition of the threat by the amygdala in his brain. This alarm then alerted the hypothalamus, the control centre in his brain for the second system, the endocrine (hormone) system.

The hypothalamus then rapidly communicated to the rest of his body through the unconscious firing of specific nerves (collectively called the sympathetic nervous system); this prepared his body for its life-saving sprint across the savannah by producing the hormones adrenaline and cortisol. His brain also stopped the perception of pain in the short-term, as it decided that violent movement (whether fighting or fleeing) was more important than recognizing tissue damage until the threat had completely gone. The third, and slightly slower, system is the immune system. Once tissue damage occurs, our immune system goes into overdrive, causing inflammation. Injury results in the release of inflammatory molecules, which recruit immune cells to fight off potential pathogens. But these inflammatory molecules also amplify pain. They increase the sensitivity of nociceptors, lowering the threshold for danger signals to be sent to the brain. This allodynia (feeling pain in response to stimuli that don't usually cause pain, such as touch) reminds us to protect the wound and not to mess with the injured area while it heals. This is called peripheral sensitization. We've all experienced it, whether when holding a pen with fingers you burned while cooking, or limping to avoid hurting a healing broken toe. Interestingly, nociceptive nerve endings release a number of inflammatory substances themselves, the best known being 'substance P'.[15] Substance P causes 'mast cells' – the skin's equivalent of landmines – to release potent inflammatory molecules that increase the diameter and permeability of skin blood vessels to allow cells from the body's immune system to come to the scene as quickly as possible. The immune system and the pain system amplify each other. This ongoing loop of inflammatory molecules sensitizing nerves, which then contributes to the inflammation, is the reason why tissue can be so sore for days and weeks after an injury.

If you want to see this interaction between the nervous and immune systems, experiment on yourself: scratch the skin along the back of your hand with a fingernail or a pointed object such as

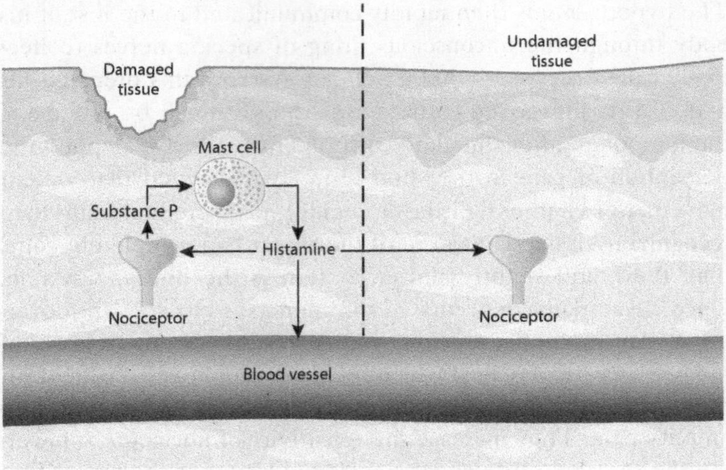

Pain and peripheral inflammation

a pencil. Three things will always happen. Firstly, within a few seconds a red line appears. This is caused by mast cells emptying out their potent contents at the site, with histamine dilating the small blood vessels of the skin to increase blood flow to the area. Secondly, after a minute or so, the redness appears to spread outside the margins of the line. This is called the axon reflex, in which the histamine activates nerve endings, which send an impulse to the spine and back to the skin, which in turn dilates more dermal blood vessels immediately around the scratch site. Finally, a weal appears along the original red line. This is because the dilation of all of these blood vessels increases their permeability, releasing blood plasma (the fluid that holds blood cells in suspension) from the vessels into the surrounding tissues. This causes the swelling that almost always accompanies inflammation. This inflammatory response is critical in our fight against injury and infection, in widening the roads to the affected area.

Short-term inflammation is an imprecise but effective form of defence and it's essential for survival and tissue healing. We should be grateful for it.

Imagine our protective mechanisms as a three-pronged trident: the three spear-tips being the nervous system, the endocrine (hormonal) system and the immune system. The shaft that connects and drives these prongs is stress: a biological response to an external threat. In short-term stress, all three of these prongs work in tandem for the single purpose of protection. All three protective systems communicate with each other in dazzlingly complex and largely undiscovered ways. But, just as short-term pain is good and protective – and long-term pain is overprotective – stress, too, is only beneficial in the short term.

Unfortunately we now live in a stressed-out society, and this is undoubtedly a cause of the growing pain pandemic. In the West the threat of violent death has largely gone, but we have replaced this with psychological stressors; we live in a slow-drip state of anxiety and pressure. We can partially blame our overly anxious ancestors for this – the ones who survived and passed on their genes were the overly alert, who assumed rocks were rhinos and sticks were snakes until proven otherwise. Stress in the modern world is not a wrecking ball but a chisel, persistently chipping away at our physical and psychological health. And it's the persistence of these modern stressors that makes them so damaging; unlock your smartphone and out jumps the judgement of social media, advertising designed to make you feel insecure and a selection of the most depressing news stories from around the world. Above you hangs a dark cloud laden with the background stressors of rent or a mortgage, bills, paperwork, job insecurity . . . ad infinitum. Our body is held in a constant state of trying to protect itself, from the emotional level to the molecular, and, as we know, this is the perfect recipe for pain.

To work out how we can start to get out of this cycle, we must take a closer look into how an over-alert immune system increases pain. We've all experienced the effects of inflammation on pain in the short term: a wound continuing to feel sore while it heals; our skin and soft tissues becoming more sensitive to pain when we go down with the flu. There's a reason why some of the oldest (and best, in certain situations) pain-relieving drugs are anti-inflammatories, which inhibit the production of various elements of the inflammatory soup. Many autoimmune and inflammatory diseases cause terrible pain – from rheumatoid arthritis to Crohn's disease – and new, targeted therapies that dampen down their chronic inflammation have revolutionized treatment for the pain experienced with these conditions. It's clear that inflammation causes pain, but it's also becoming evident that long-term inflammation may underlie the development of persistent pain. For instance, there is evidence that chronic, low-grade inflammation of the brain – neuroinflammation – contributes to persistent pain and the memory deficits that commonly accompany it.[16] We've all experienced short-term neuroinflammation: it's responsible for 'sickness behaviour', that horrible feeling of fatigue and depression when we get an infection. This is a completely normal response to short-term infection and may play a role in us withdrawing from others so as not to spread the infection. But research over the past few decades is building a picture of chronic, low-grade inflammation causing or worsening a number of neurological and psychiatric conditions. This inflammation in the brain may be much more common than previously thought. It can even reduce neuroplasticity, our brain's way of adapting itself.[17] And inflammation has a long memory.

Immunity – a word that is now common parlance following the Covid-19 pandemic – is a state where our immune system recognizes pathogens that have entered our body before and can mount a powerful, targeted immune response with antibodies and T-cells. But there is a simpler, more primitive element of the immune system that can remember infection or injury, and this can worsen pain. Pattern recognition receptors (PRRs) are protein receptors, found on immune cells (responsible for the initial activation of the immune response), that recognize generic molecules common to most pathogens (known as pathogen-associated molecular patterns, or PAMPs) and generic molecules released when body tissue is damaged (known as damage-associated molecular patterns, or DAMPs). PRRs are essentially barcode readers. When they detect PAMP or DAMP patterns, they instruct their cells to release inflammatory molecules, which brings about a protective response: increasing pain and recruiting immune cells to the scene. They're good at recognizing anything that is foreign to the body but they're pretty indiscriminate, like a guard dog that tries to attack a friendly postman as well as a thief. Fascinatingly, we now know that opioid medications activate these receptors. Opioids, such as morphine, are wonderful at dampening short-term pain, but research also clearly shows that while they dial down nociceptors, they also activate these inflammatory PRRs. Over time, their anti-nociceptive effect goes down, while still stimulating an inflammatory response, which increases pain. This is why, over a long period of time, opioids can actually make pain worse. This opioid-induced hyperalgesia (increased pain) is widely known, but it's only as recently as 2019 that the immune system has been exposed as the culprit.[18]

It is becoming clear that long-term inflammation 'primes' the immune system, increasing the amount of inflammation released in any new response. It is also increasingly evident that psychological stress and fear increase our peripheral immune response, making us more sensitive to pain. Psychological stress alone increases inflammation in the skin and body for days, probably because it is preparing to fight the infections that come with the blow of a weapon or the bite of an animal – or, in Henry's case, a foot-long thorn of hardened protein.[19] It is amazing to think that even a simple thought – the retrieval of a traumatic memory or anxiety about things that could yet happen – could create inflammation.[20] In a study where mice were taken out of their home and given electric shocks in a certain area of the lab, they associated that area with stress and pain. If they were brought back to that place, levels of their inflammatory molecules increased. Interestingly, if they were then given shocks in their comfortable home, these molecules did not increase.[21] When we learn fear, we create inflammation. This is probably why we start to sweat when returning to a school we hated, or why spending time with a family member we feel uncomfortable around leaves us with knots in our stomach or a throbbing headache. Like pain, inflammation is good in the short term, but bad in the long. Short-term inflammation is like a country's efficient military response to invasion. Long-term stress is when the military gains too much power and begins to turn the country into a paranoid, hypervigilant police state.

Chronic inflammation is worsened by a number of things. For starters, it increases with time. In fact, inflammation is the main driver behind biological ageing, leading to the misleadingly cute moniker 'inflamm-aging'. Inflammation thickens our arteries, accelerates the onset of dementia and, of course, makes ageing painful. It is the Faustian pact our body makes in having an immune army; it is the slow, collateral damage of old microbial wars. But there are many factors that we can control to reduce inflammation and quieten its associated pain. Chronic psychological stress and anxiety are other powerful drivers of inflammation.[22] This is miserably seen in the strong link between adverse events in early life and both inflammation and chronic pain throughout later life.[23] If we don't look after our mental health and seek help if needed, we won't even begin to look after our physical health. Relaxation calms down inflammation, and it's likely that this explains how meditation, tai chi and yoga are often effective pain relievers.[24] Long-term inflammation is also strongly associated with social isolation; once again, our biology is imploring us not to ignore the social elements of disease.[25]

Another contributor to inflammation and pain is the constellation of risk factors known as 'metabolic syndrome': obesity, a sedentary lifestyle, high blood pressure, high blood sugar and high cholesterol. It is important to state that carrying more fat produces pain not just because it increases weight on joints, but also (and probably more so) because fat itself is pro-inflammatory.[26] The solution is wonderfully cheap and low-tech: keep moving and eat sensibly! There are many 'anti-inflammatory' diets out there, all built on wildly differing foundations of evidence – the science of nutrition is labyrinthine in its complexity – but the general message is clear: have a balanced diet that includes lots of plants and fibre (these encourage the growth of diverse populations of gut microbes), eat foods rich in omega-3 and lose excess fat in a sustainable way. The key, then, is not to anxiously pursue a rigid anti-inflammatory diet, but to avoid a pro-inflammatory one: overeating. It also almost goes without saying that smoking is pro-inflammatory, and smokers are three times more likely to be in chronic pain than non-smokers.[27] The evidence for quitting is overwhelming. Cutting down on alcohol and being sensible with caffeine are other powerful but overlooked ways of reducing stress, inflammation and pain.

Until fairly recently I had completely ignored an activity that powerfully influences persistent pain. But in 2020 I wrote an essay and presented a talk to the Royal Society of Medicine on the relationship between pain and an activity we spend a third of our lives engaged in: sleep. In my research for the essay, it was patently easy to see how closely they were linked: roughly three quarters of persistent-pain sufferers experience sleep disturbance and half of individuals with insomnia have persistent pain.[28,29] There's some debate about the chicken-and-egg nature of the relationship: does pain lead to bad sleep or does bad sleep worsen pain? The short answer is both – it's bidirectional. Intriguingly, evidence suggests that insomnia could be the greater driver of this cycle; insomnia causes pain more than pain causes insomnia.[30] In any case, it's a bad predicament. A 2019 study – surprisingly, one of only the first brain-imaging studies we have on sleep deprivation and pain – shows that insomnia amplifies pain in a two-pronged attack. It increases pain reactivity in the somatosensory cortex (the area of the brain generally responsible for detecting the type of danger signal, as well as its location on the body) and reduces the ability of the decision-making parts of our brain to moderate pain.[31] The study found that even mild reductions in sleep worsened pain. Importantly, sleep disturbance is also a one-way ticket to over-inflammation. Even one night of sleep loss causes inflammation in the body.[32] Sleep disturbance is also on the rise in our stressed-out society; since the invention of the light bulb, we sleep for one hour less per day. The internet has reduced that by another half-hour.[33]

My first glimpse of the vicious pain–sleep cycle was as a first-year junior doctor on an elderly-care ward. After the morning ward-round I was to take a blood sample from Mary, an eighty-year-old woman admitted with pneumonia. The pneumonia had now resolved, but her worsening persistent back pain – seemingly unaffected by regular opioids – was posing the greatest challenge with regards to discharging her safely and sustainably into the community. While starting to apply the tourniquet, I initiated some needle-distracting small talk.

'Did you sleep well last night, Mary?'

'No, I never get any sleep, love! At least, not at night. Since having this pain, I've given up on getting a good night's sleep.'

At the time, I didn't realize that her words revealed insights into a complex downward spiral that was taking a terrible physical, psychological and social toll on her. Over the next few days I would see Mary napping intermittently throughout the day. This not only reduced sleep pressure (the feeling of sleepiness) and hindered good-quality sleep at night, but also robbed her of opportunities to receive the vital analgesia of movement, whether with physiotherapists and occupational therapists or simply walking (with assistance) to the activity room.

Looking back, and examining the available research, I've

realized that there is another factor that was worsening Mary's downward spiral. Opioid drugs, in the long term, reduce sleep-quality and worsen sleep-disordered breathing, which can cause or worsen pain.[34] This is another way in which opioids can worsen the very thing they are meant to treat. Opioids also stimulate arousal nuclei in the brain, which has a similar effect to caffeine.[35] Insomnia hurts, and opioids often make this worse. To tackle persistent pain, we need to take sleep deprivation seriously. As I unintentionally discovered when asking Mary about how she was sleeping, solving this problem in every persistent-pain sufferer could start with a general practitioner (or any other healthcare professional, for that matter) conducting a simple sleep assessment. This conversation could put into motion a hopeful journey out of a destructive, self-perpetuating cycle and enable the individual to manage and live with their persistent pain.

Perhaps all of these complex, fuzzy 'lifestyle factors' should be summarized by a pain parable. Our body is like a beautiful, intricate garden. Persistent pain is like a thorny weed. It feeds on soil – unchangeable factors such as past tissue damage, past trauma, upbringing and genetics. But to grow it also needs to be watered – with stress and inflammation: psychological stress, smoking, a poor diet, insomnia, a lack of exercise, anxiety and social isolation, to name a few. Fortunately, we have control over how much of this we allow to water the weed, albeit to different extents depending on our circumstances. I often see doctors exhibiting what I call the cynic reflex: a rolling of the eyeballs in response to hearing the word 'holistic'. But we are all wonderfully unique individuals living in a unique social and physical environment; we need to be seen as a whole. Understandably, people living with pain want direct answers for their horrible experience, but in most cases of persistent pain it is much more effective to address these vulnerabilities than to chase a specific cause of the pain. If we focus on changing one of these lifestyle factors at a time, we can start to move out of the downward spiral of pain and begin to live again. And what could go right? The only side effect of reducing stress and inflammation is improved health.

Pain is all about protection. Reducing stress and increasing a sense of safety is something we should strive for at the individual and societal level. But to truly understand how to change pain, we need to peel off another layer of the onion and find out how we get from short-term to persistent pain in the first place. This lies in the exciting new science of our changeable brain.

11

The Runaway Brain

Why pain remains

Any man could, if he were so inclined, be the sculptor of his own brain.

SANTIAGO RAMÓN Y CAJAL, NOBEL PRIZE-WINNING NEUROSCIENTIST

FOR AS LONG as I can remember, my local supermarket has muddled through with an overly enthusiastic anti-theft alarm at its exit. The alarm gates are designed to release radio waves from a transmitter in one side of the gate, which is detected by tiny antennae on the labels of goods. These labels then transmit their own signal at a specific frequency, which is picked up by a receiver on the other side of the gate. If the checkout assistant hasn't passed over the label with a deactivating device (or, of course, you have stolen the item), the gate detects the signal and emits a loud, attention-grabbing beep. Pretty clever. For some reason, my local supermarket's gates decide to beep every other person that passes through them, regardless of whether they have even bought anything or not. Instead of replacing the gates, the shop has an unfortunate member of staff permanently stationed next to them, sheepishly apologizing to the innocent shoppers impolitely beeped by the gates, like the owner of an overly aggressive dog. I like to think – not least because it helps with the metaphor – that once upon a time the shop's management noticed that a few of their finest Dom Pérignons had gone missing, so they decided to turn up the sensitivity of the gates to catch the miscreants. But even though the thief has now left the area or has renounced their criminal ways, the oversensitive gates now interpret any movement as a stolen item.

Persistent pain works in a similar way. There was once a reason for the sensors to go off – a twinge in a lower back muscle, say – but the brain, fearful of any damage to the spine, interprets any movement in the lower back as danger and causes pain even long after the initial tissue damage has completely healed. The longer the pain goes on for, the better the brain gets at making pain and the weaker the connection between pain and tissue damage. In most cases of persistent pain, the brain is receiving and interpreting information that now no longer reflects what is going on in the tissues. It is a false alarm. Pain has developed an afterlife.

A key concept that bridges short-term and long-term pain is something called 'central sensitization', first described by the eminent South African neuroscientist Clifford Woolf in the 1980s.[1] The word 'central' simply refers to the central nervous system: our brain and spinal cord. Sensitization is the process by which these structures become hypersensitive: you end up getting more pain with less provocation. Pain also takes longer to disappear, if at all: it's as though the volume knob for danger is turned up and gets jammed there, increasing the gain on pain. This manifests itself in different ways. We might suffer allodynia – pain in response to sensations that are usually non-painful (such as touch). We experience this if we get sunburned: even the most casual of touches can feel like a stinging slap and a warm shower becomes a searing torrent of lava. Another is hyperalgesia – feeling excessive pain in response to noxious stimuli. If you stub your toe on a doorframe, it hurts. If you stub it again on the way back, while it is still tender, the same force on the same object produces much more pain. Central sensitization can also make pain take longer to disappear after exposure to a stimulus.

To see how central sensitization occurs, let's say you twinge a muscle in your lower back. Nociceptors – the body's danger detectors – are activated and transmit a signal from the mildly damaged muscle, along the peripheral sensory nerve, towards the spinal cord. We saw in Chapter 1 that the danger signal has to travel from the peripheral nerve to a spinal nerve, as this is what takes the signal up to the brain. The connection between the nerves is a microscopic space called a synapse. Once the electrical danger signal reaches this synapse in the spinal cord, chemicals called neurotransmitters are released into the synapse, which either excite or inhibit the next nerve. In varying and complex ways – such as changes in levels of receptors and transmitters in the spinal cord – the more intensely or longer these secondary nerves are activated, the bigger the response to the same stimulus becomes.[2] This is why, even when the small amount of tissue damage from the muscle-twinge has completely healed, a pain 'memory' exists in the spine. The more you feel the pain and the more easily your brain recognizes it, the more intense it feels – this is termed 'pain wind-up'. And this is just a small snapshot of what's going on in the spine and brain. In innumerable levels of complexity, impulses from the brain can travel down to the spinal cord and either increase or decrease the power of the synapse. And synapses are not just in the nerves of the body and the spine: the brain has around 100 billion nerves, and possibly up to 10,000 billion connections. This is how, through the strengthening and weakening of different synapses in different pathways, we can 'rewire' our brains – the concept that 'neurons that fire together, wire together'. The more we use certain brain circuits, the stronger they become. Conversely, the less we activate them, the weaker they get – 'use it or lose it'.

Think of the circuits and networks in the brain as paths in a forest. I love going for walks and runs in new areas of countryside and I tend to choose a well-worn path used by walkers who have gone before me. In running on this path, I – ever so slightly – further consolidate and widen it. If walkers and runners started to use another route to traverse the forest, the well-worn path would gradually weaken and, maybe – once enough undergrowth had sprung up over it – disappear completely. Our brain is similar: paths and networks activated frequently become stronger, and those that get forgotten or become underused fade into insignificance. The brain is wonderfully 'neuroplastic', capable of immense change and flexibility. We are constantly rewiring our brain, for good and ill – when we learn someone's name and associate it with their appearance, when we practise a new musical instrument, train in a sport or develop a habit. I'm slightly uncomfortable describing persistent pain as a 'habit', as it has false connotations that you bring the pain upon yourself. However, in many ways it describes what persistent pain is: a pattern of feelings and thoughts stuck in a fixed loop, a vicious cycle of pain and hyper-arousal. It etches pain on the brain that can stay there well after the cause of the initial pain has gone and the injury has healed. Central sensitization and its rewiring of the brain is complex and unique for each individual, but it is simple in concept: it is a protective brain's overreaction to threat.

Returning to the issue of your back, when the danger signal first reaches your brain from the muscle in your back, the brain decides to create pain. As this pain is new, sharp and pretty severe, you wonder, *What could this be? A slipped disc? A pinched nerve? Have I damaged my spine?* The fear that this could be severe or permanent damage makes you hyper-attentive to the signals travelling from this area of damaged muscle, piling on stress and making the experience of pain worse. Whenever you walk or stand up, you are anticipating the twinge. Your brain begins to interpret normal muscular movement in your back as painful. Your belief that your lower back is vulnerable and needs protecting actually strengthens the pain circuits, and in a vicious spiral this worsens your overall view of your back as well (*My back is crumbling!*). Whenever you, or someone else, affirms how bad your back has become, these circuits in the brain, these negative neurosignatures, are strengthened. With depressing irony, the false alarms of persistent pain drive us to behaviours that genuinely worsen the pain. After an injury, it is sensible to rest for a short period of time, but there is an overwhelming amount of evidence demonstrating that long-term immobility worsens pain and is a disaster for all of our bodily systems. As we have seen, the pain begins to rob you of your sleep, your mood plummets and your stress rises, all of which worsens pain. Pain causes us to withdraw from danger, but persistent pain makes us withdraw from life; we stop going out to seek purpose and enjoyment, and our social support network dwindles. Down and down the spiral you fall, and the signature of 'PAIN' becomes ever-more deeply engraved on your brain. Persistent pain is, in many senses, a 'learned' pain.

In 2018, a team led by Tor Wager at the University of Colorado Boulder found another piece in the puzzle.[3] In a study, they first explained to a group of participants that they would be placed in an fMRI scanner, shown visual cues – either the word 'low' or 'high' – and then be given a blast of low or high heat respectively. In fact, there was no relationship between the cue and heat-intensity. However, when the participants saw the word 'high' they expected more heat than if they saw 'low', and felt more pain, regardless of how hot the stimulus actually was. Our expectations powerfully influence our perception of pain, but what was really interesting was how easily the participants 'learned' pain. If the participant expected a high level of pain and received it, they ended up experiencing more pain the next time they received the same stimulus. However, if they expected a high level of pain but received a lower stimulus, their perception of pain didn't diminish but instead stayed the same. In an amplifying spiral – a self-fulfilling prophecy – the more pain we expect, the more our brain creates, which makes us expect more pain the next time, and so on, and so on. We humans are highly susceptible to confirmation bias – favouring information that reinforces our own beliefs – and when it comes to survival, our brain can be a real pessimist. It's easier to climb up the pain ladder than climb down, and it often stops us realizing that our tissues are healing and that we are getting better.

At this point you would be completely at liberty to ask, 'So, are you saying that it's all in my head? Are you dismissing the horrible, real pain I feel in a specific part of my body as psychological?' No, and no. Persistent pain certainly is real. And while psychology certainly influences pain (just as it influences many, many diseases), pain is best seen as neurological – a disease of nerves and brain circuitry. It's no less real than epilepsy.

There's also a mountain of research supporting real brain changes in persistent pain. Around a fifth of people who have knee replacements find that it doesn't reduce their pain at all. In 2019, a team at the University of Oxford analysed these patients' brains using fMRI imaging and found that this subset of patients had a different, 'chronic pain', neurosignature compared with those whose pain had been relieved by the surgery. The neuroimaging suggested that there was both an increased facilitation of danger signals travelling up from their spinal cord and reduced descending inhibitory signals from their brains. The mechanical problem in the knee had been sorted, but pain had been branded on the brain.[4]

Other studies show that persistent pain changes patterns in the brain even when the individual is not currently experiencing pain.[5] Most worrying of all, there is evidence that persistent pain ages the brain. One 2004 study compared the density of grey matter in the brain between people with persistent pain and healthy controls. It found that if persistent pain lasts more than five years, it can shrink the grey matter of the brain by 5–11 per cent, which is the equivalent of ten to twenty years of normal ageing.[6] Many other strands of research support the concept of persistent pain as a case of neuroplasticity gone wrong. One strand of particular note in recent years is that of genetics. A 2019 study that quarried the rich data mine of the UK Biobank identified seventy-six genes to be risk factors for persistent pain, many of which encode key processes of neuroplasticity.[7] Another paper from 2019 identified a mutation in a gene, linked to persistent pain, that controls an enzyme responsible for normal serotonin levels.[8] People with this mutation have lower-than-normal serotonin levels as a consequence, which leads to increased 'somatic awareness' – an increased awareness of all kinds of bodily sensations. This is wonderful when it comes to pleasurable sensations, but in the face of pain it is fuel for anxiety, a sense of threat and an amplification of pain. Remarkably, this mutation is not rare: it's present in about 10 per cent of the population.

At this point we need to return to a crucial caveat. In most cases of persistent pain, whatever caused the initial injury has healed. Pain is now the primary disease. But there are a number of cases where there is continual damage that triggers nociceptive fibres; chronic inflammatory diseases are good examples. It is also important to point out that not *every* case of back pain is our brain's overreaction. A small – but important – minority of cases are caused by serious conditions – cancer, some infections, spinal fractures and the nerve-compressing cauda equina syndrome – but these can usually be ruled out by doctors, who will be on the lookout for 'red flag' symptoms. However, in the majority of cases of persistent pain (and over 90 per cent of cases of back pain), there is no longer any identifiable tissue damage; our brain has become hypersensitive.

Perhaps the starkest example of central sensitization is fibromyalgia. This is a real and often devastating condition of widespread muscle pain, increased pain sensitivity, fatigue and the infamous 'fibrofog': a disruption in memory and mental processing. Fibromyalgia is particularly frustrating because there is no medical or scientific consensus as to its cause, and most doctors have a poor general understanding of it. Doctors are never more uncomfortable than when faced with unexplained symptoms and an unexplainable disease, particularly ones that can't be measured with a test and treated with a course of medication or an operation. When they don't have the answers, some health professionals get defensive, dismissive and even discriminatory – 'fibro faker' and 'inadequate woman' are just two indefensible phrases I have heard from the mouths of qualified doctors. I often see patients with fibromyalgia passed between primary care physicians, rheumatologists and neurologists like a hot potato. An emerging literature, however, is showing that a core mechanism in most cases of fibromyalgia is central sensitization, when the brain's alarm system of pain becomes overactive.[9,10] This helps prove that fibromyalgia is real, but – and this is an important 'but' – we still don't know what causes the sensitization. As is likely to be the case in persistent pain generally, my money is on the immune system. In 2017, Scandinavian researchers analysed inflammatory markers in people with fibromyalgia, finding evidence for inflammation in the brain (neuroinflammation) and the body (systemic inflammation).[11] In 2019, teams at Harvard and the Karolinska Institutet in Sweden found that microglia – immune cells of the brain – are highly activated in fibromyalgia patients compared with healthy controls.[12] While the science is young, it is very exciting: neuroinflammation may prove to be a significant driver of central sensitization and of persistent pain in general.

Millions of people across the world have experienced the transition from acute to persistent pain. But to gain a full understanding of our changeable, neuroplastic brain – how it gets us into pain and how it can get us out of it – we have to look at an uncommon and odd condition, one of the most mysterious and magical aspects of the human experience. During a visit to a hospital in a remote corner of north-east India, I came across a patient who felt pain in the air. Ten years earlier, Aman had been driving his brightly coloured truck on the ten-hour journey from the plains of Assam up into the jungle-clad foothills of the Himalayas, somewhere near the Burmese border. The steep, winding roads carved into the mountainside were little more than mud tracks even on a good day, and it is a terrifying, vertiginous journey I wish never to repeat; Aman had been trying to make the ascent in a full-blown monsoon. About halfway into his ascent, the mountainside above him suddenly collapsed in a slump of mud and rock, slamming into his orange truck and sending him careering off the mountainside. The vehicle fortuitously smashed against a clump of thick trees sticking out of the cliff a few metres below – otherwise he would have fallen to a certain death. Aman's right arm took the full force of the landing, however, with his forearm and elbow completely crushed. After a lengthy rescue from this precipitous position, the local doctors successfully carried out a below-shoulder amputation in the hospital.

As Aman and I talked, his face intermittently screwed up into a fleeting grimace. He explained that several times each day he got the sensation that his whole right arm was still there, accompanied by the feeling that his invisible fingers were being scalded with boiling water. Phantom limb pain is remarkable: the sensation of pain in a hand that is not there clearly demonstrates that pain is created in the brain. It is also perhaps the most powerful image of persistent pain, a case of neuroplasticity gone wrong. You would be forgiven for thinking that this condition is a curiosity on the fringes of medicine, but it's surprisingly common: over three quarters of amputees experience phantom limb pain.[13] After medical school and one year in as a junior doctor, I had already seen hundreds of cases of phantom limb pain, from soldiers who had lost limbs in IED explosions on Afghan roads to elective amputations due to peripheral vascular disease. Some symptoms were bizarre, from someone feeling their phantom hand waving at passing nurses to another feeling their phantom leg gradually 'telescope' – becoming shorter and shorter. These ghost pains have mystified medics for centuries, creating a smorgasbord of theories to explain them. Admiral Nelson, the British naval hero, lost his right arm in the Battle of Santa Cruz de Tenerife. He argued that his subsequent phantom limb pain was 'direct evidence for the existence of the soul'.[14] Throughout medical school, and long before I looked into the science of phantom limbs, I assumed that the cause of this bizarre condition was damaged nerve endings in the stump sending aberrant signals up to the brain. Until the last decades of the twentieth century, this was actually one of the main schools of thought. Surgeons would try to exorcize the phantoms by performing amputations further and further up the limbs in an attempt to remove the damaged 'pain receptors'. But the phantoms always came back, and often with a vengeance.

V. S. Ramachandran, a distinguished and fiercely inventive Indian–American neuroscientist, had been fascinated by phantom limbs since medical school. In the early 1990s, he began to wonder whether 'phantom limb syndrome' was caused by neuroplasticity. He built this theory on the sensory maps in our brain discovered by an outstanding Canadian neurosurgeon, Dr Wilder Penfield. Penfield spent the 1950s dedicated to treating patients with intractable epilepsy.[15] Many of his patients suffering with epilepsy tended to experience auras – a sense that a seizure was about to start. He theorized that if, having removed a piece of the patient's skull, he could arouse the aura by touching an area of their brain with an electrode while they were fully conscious, he could find the area of brain responsible for the seizure. While this experiment was only moderately successful, he stumbled upon something even more remarkable. When Penfield prodded different parts of the surface of the brain during surgery, patients felt sensations in different parts of their skin. Penfield then painstakingly recorded which areas of the brain corresponded with sensations on different areas of skin. Interestingly, the brain's 'body map' of the sensory skin appears jumbled up: the area of the brain that represents the toes is right next to the genitals, and the hand area is right next to the face area. Also, the size of the area of brain used is not related to the amount of the body's surface area that the skin covers. For example, the skin on the fingertip of our forefinger requires the attention of a proportionally much larger area of our brain's body map than, say, the skin on our back, which isn't surprising given the considerable density of sensory receptors in our forefinger. To represent this, Penfield created a model of a human where the appendages of the body were either shrunken or enlarged to represent the amount of space they occupy in the brain. The result was what he named the sensory homunculus – a gangly, off-balance, 'grotesque creature' (as Penfield described it) with enlarged body parts where there is a high density of receptors (hands, feet and lips), in contrast with those where there are far fewer (such as the torso and arms), which are spindly and diminutive.

Ramachandran speculated that phantom limbs were not caused by damage to nerves in the stump, but instead resulted from a rewiring of the brain, a redrawing of the sensory homunculus map. To test this, he assessed a young man – let's call him Mike – who had a phantom itch in his missing left hand. The man's arm had been amputated just above the elbow following a horrible car accident in which he was thrown out of the car and – while still in the air – could see his severed left hand gripping the steering wheel. Ramachandran used a cotton bud to touch areas of the young man's skin and asked him what he felt. Nothing exciting happened until he reached Mike's cheek: he could feel the sensation in his phantom hand. Further exploration revealed that specific areas of his face corresponded to areas of the missing hand; stroking his upper lip would be felt on his phantom index finger, for example. Amazingly, Mike could now finally scratch the itch in his phantom hand by simply scratching his cheek. This makes sense in light of neuroplasticity and how the body is represented in our brain's map: when the hand was lost, connections sprouted from the closest brain area to the part of the brain for the missing hand – the face – to make use of this 'dead space'. This could even explain foot fetishes: in the brain's map, our toes are placed right next to our genitals. Many lower-limb amputees find the end of their stump to be sexually stimulating, as the brain's area representing the genitals begins to make use of this new space. Some amputees even feel the sensation of urination over the stump.

Ramachandran then used neuroimaging to demonstrate that Mike's brain activity had migrated from the hand area of his brain's map to the face area.[16] Other studies show that those amputees who do not go on to experience phantom limb pain do not exhibit this 'spillage' and takeover of the newly vacant area.[17] A seminal study by Professor Herta Flor, a leading pain scientist based at the University of Heidelberg, Germany, found that the level of phantom pain was proportional to the level of this neurological reorganization.[18] Phantom limb pain really is a case of neuroplasticity gone wrong. Over the following years, Ramachandran went on to assess hundreds of amputees and he began to notice a pattern. Many felt that their phantom limb had 'frozen', feeling as though their missing limb was present but rigidly stuck in a fixed position. He began to realize that these were the people whose limbs had been immobile in slings or casts leading up to the amputation. It seemed their brain had added the immobile limb to its body map, and this had persisted after the limb was removed. Ramachandran was slowly excavating some remarkable facts about the brain, and it was becoming very evident that the brain – and the way it creates our perception of our body – is dynamic and changeable.

What can only be called a stroke of genius led to his creation of an astounding device – based on a hypothesis that phantom limb pain is 'remembered' pain – that has helped a great many amputees 'forget' that pain. Imagine a medium-sized cuboid box, roughly the size a new toaster would arrive in. The box is open at the top and has two arm-shaped holes at the front. A divider separates two compartments of the box. One side of the divider is a mirror and it faces the compartment the patient puts their non-amputated arm into. If the patient looks down at the box and leans slightly towards their non-amputated arm, it appears as though their missing arm has reappeared in the other compartment. The mirror box is a simple illusion that often has seemingly miraculous effects. When the patient moves their good arm, they can 'see' their previously painful, frozen, phantom arm moving. When Ramachandran gave patients the box to take home and practise with, he found something even more remarkable. Many reported that the phantom pain had completely disappeared. He understood that pain is, essentially, an illusion – or, at least, not a completely accurate representation of reality. And so he treated it with an illusion; he had performed the first phantom limb 'amputation' in history. The very sight of a phantom limb moving can be enough to cure someone of phantom limb pain; the very process of thinking that you are better actually makes you better. It's crazy, but true. Ramachandran's discoveries take us to the core of what pain is – and what it isn't. He puts it wonderfully in his book, *Phantoms in the Brain*:

> 'Pain is an opinion on the organism's state of health rather than a mere reflective response to an injury. There is no direct hotline from pain receptors to "pain centres" in the brain. There is so much interaction between different brain centres, like those concerned with vision and touch, that even the mere visual appearance of an opening fist can actually feed all the way back into the patient's motor and touch pathways, allowing him to feel the fist opening, thereby killing an illusory pain in a non-existent hand.'[19]

We feel pain where our brain thinks the damage is, which is not always where it actually is. Pain, therefore, is created in the brain and projected onto the body. We all have phantom limbs; it's just that our physical body gets in the way. We can't distinguish our body from our brain's projection of it. Ramachandran argues that our 'whole body is a phantom, one that your brain has constructed purely for convenience'.[20] Our 'body image' is formed in the brain and projected onto the body. Crucially, it is independent of our physical body. Phantom limb pain is, in a sense, a condition of distorted body image. This also helps explain something very commonly seen in persistent pain: spreading pain. As the brain becomes hypersensitive to pain through central sensitization, a pain map is created in the brain and this often spills over to other areas of the brain. We can end up being hurt in one part of the body but feeling the pain in another, known as referred pain. Persistent pain caused by an initial injury to one area can also spread over the whole body: one 2009 study found that patients who had persistent neck pain following a whiplash injury were more sensitive to pain in all areas of their body, most of which were never injured.[21]

Ramachandran's mirror-box therapy is very effective for many people suffering with phantom limb pain,[22] but is also a cure for some people with certain forms of persistent pain. One 2003 study found that when people who had one arm debilitated by 'complex regional pain syndrome' (CRPS) – when a region of the body is in extreme persistent pain, disproportionate to an initial injury – moved their healthy arm in the box to create the illusion of their affected arm moving painlessly, the pain gradually lessened.[23] The process of thinking that the arm was pain-free was enough to gradually rewire the brain into making it pain-free. It's also likely that seeing the illusory hand move diminishes the association our brain makes between a particular movement and a specific pain. The feeling of pain is almost always accompanied by a motor action (movement): we retract our hand from a hot plate; we hold and guard an injured wrist. Seeing the motor action without the associated pain loosens the grip of pain in the brain.

But the mirror box was effective only for those who had suffered persistent pain for just a few months; perhaps those who'd had the pain for longer had brain circuitry wired in such a hyperprotective way that the brain forgot how to move the limb without pain. In a case of overprotection gone wild, the brain has made the limb 'guarded', so that any movement whatsoever brings about extreme pain. This is where the Australian pain scientist (and pain explainer extraordinaire) Professor Lorimer Moseley comes in. He devised a rehabilitation programme called 'graded motor imagery' (GMI), which gradually exposes people to movement, 'sneaking under the pain radar', as he calls it. In a nutshell, GMI consists of three phases. The first involves viewing images of body parts and identifying, as quickly as possible, whether they are on the left- or right-hand side of someone else's body. Individuals with persistent pain tend to become worse at discriminating areas of their brain's body map, so this training helps lay the foundation for healing neuroplasticity. In the second stage, the participant has to imagine certain movements without actually moving, thereby stimulating and training the same areas of the brain that are activated in movement, but sneaking under the brain's pain radar. Finally, they undergo mirror therapy. Rigorous studies show that graded motor imagery is effective for very persistent CRPS and phantom pain,[24] and there's no reason why it wouldn't be effective in many other types of persistent pain. And while we must admire the low-tech illusory genius of mirrors, more advanced technologies are also getting in on the act. In 2018, teams at University College London and the University of Oxford used non-invasive 'transcranial direct current stimulation' (where low electric currents are delivered through electrodes attached to the head) in amputees to stimulate the hand area of their brain map while they tried to imagine moving their phantom hand, with 30–50 per cent reductions in pain.[25] And, of course, if mirrors are effective at updating the brain's body image to one of safety, health and reduced pain, then our new illusory technologies should be very successful. Recent studies suggest that virtual reality (VR) sessions that recreate the amputee's missing limb in an immersive and enjoyable environment are also immensely effective at relieving pain.[26]

It is undeniable that our brain is wonderfully changeable – fantastically plastic – throughout our life. Our brain helps us learn, grow and love. It also wants to protect us, and it can sometimes do that job too well. Persistent pain is, in most cases, the symptom of a runaway brain. But while most of this chapter has been dedicated to the doom and gloom of neuroplasticity's ability to cause pain, we are beginning to realize that the same mechanism can get us out of it. If we can change the brain, we can change the pain.

12

The Pain Revolution

A new hope for persistent pain

Just as the source of suffering lies in humanity, the cure may be found in our humanity also.

Dr Frank Vertosick, American neurosurgeon

'When I mention the "K word", doctors aren't interested. But when I introduce my treatment as "a bilateral, rhythmic, psychosocial intervention" their ears prick up!'

Positivity radiated out of Betsan Corkhill like heat from a fire. When the Bath-based well-being coach told me about how she had come across a certain pain-healing intervention, she spoke not as someone trying to persuasively package and sell a new treatment, but as someone who had unexpectedly stumbled across a hidden chest of treasure. The treasure in question? Knitting.

Corkhill trained as a physiotherapist at The Middlesex Hospital, London, in the 1970s. While her training was vigorous and hands-on, her early years of practice in the UK were disheartening. The prevailing medical view was (and in many ways still is) one of structuralism: treating body parts and biomechanical problems rather than seeing the whole person. Her eyes were opened when she spent time working in a medical rehabilitation clinic in Switzerland. She saw that if patients were treated as individual people, with their mental and social health seen as being just as important as their physical health, they had better outcomes, and were better able to take control of their chronic disease journey. In her book

Knitting for Health and Wellness she writes: 'I learned that if people feel good about themselves, and are encouraged to be active, interested and social, they heal.'[1] When Corkhill got back to the UK, she was shocked by how the system didn't allow for the brain to learn, change and improve. The last straw was when Corkhill was called in to see a relatively young man in the community who'd had a severe stroke and was paralysed down one side. Instead of encouraging rehabilitation of his brain and his arm to relearn motor skills, the medical team had instructed his arm be strapped to an arm of his wheelchair each morning in a misguided attempt to relax his tight muscles. Now completely disillusioned, she quit.

She retrained as a production editor for a range of magazines, which included a number of craft titles. 'One of the first jobs they put me in charge of was to organize the letters pages. But as I began to read the letters it soon became evident that almost all, say 98 per cent, of these letters talked about how therapeutic people found crafting, and knitting in particular. I mentioned this to the editor and she brought me to a filing cabinet of previous letters. There were thousands. I'd stumbled across something really important. There were vast numbers of people from all over the world, from different backgrounds, different cultures, who said more or less exactly the same thing: knitting healed them. The first letter I picked up was from a fourteen-year-old girl who'd been in and out of hospital with long-term pain, and she said, "When I knit, I don't need to take my pain medications."'

The research evidence had found Corkhill; it was crying out to her. Here were thousands of testimonies, voices of people who seemed to be benefiting from something completely overlooked by mainstream medicine. She would later be involved in a comprehensive survey of over 3,500 knitters, of whom 90 per cent of those who had persistent pain said they felt that knitting was a successful means of coping with their condition.[2]

This discovery seems to be bordering on the miraculous, so I had to ask: 'How?'

'Knitting works in lots of ways, but I summarize it in a "knitting equation": knitting = movement + an enriched environment + social engagement,' she explained.

Movement, if done gently and built up gradually, is one of the best pain relievers. Movement opens up the body's drug cabinet of anti-inflammatory and analgesic molecules – dampening down danger signals travelling to the brain – and helps to heal and nourish body tissues. But the movements involved in knitting have even more layers of complexity. The rhythmic, repetitive movements stimulate the release of serotonin: a natural mood booster and pain soother.[3] This is also why people in pain or distress tap, rock or pace. Perhaps we can learn something from the stereotypical old grandmother who knits while gently swaying on her rocking chair. Learning bilateral, coordinated movements that involve concentration and rely on visual input really does rewire the brain.[4] The hand movements involved in knitting also, importantly, cross the body's vertical midline. An experiment at the University of Milano-Bicocca in Italy found that the pain stimulated by focusing a laser on the back of someone's hand is reduced if the arms are crossed and the hand is on the opposite side of the body's midline.[5] This may be due to the fact that crossing hands over the body interferes with the brain's way of localizing a dangerous stimulus on the body.

Knitting also allows someone to gently expand their personal space. This may not seem to have much relevance to persistent pain, but it really does. We all have a sense of personal space – a 'red zone' in which we feel comfortable and safe. If anyone or anything enters our personal space, we feel the need to move away from them (or get them away from us). Its boundaries are constantly moving and are mostly decided subconsciously; normally it is the space we can reach and touch within a single step, but it adapts to be much smaller in a mosh pit of a music gig or much larger if you've just found a quiet, isolated picnic spot. It's all about safety and protection, and in persistent pain, where the mind and body are on hyper-alert, the area we want to protect extends disproportionately far; we're so conscious of our painful right shoulder that any touch to the right-hand side of the body causes tension and pain. This over-guarding of our own turf makes us more likely to see things out in the world as being a potential threat, reducing our drive to move about and explore, which in turn makes the pain worse. Knitting encourages people to slowly feel their way outwards, gaining a sense of safety and encouragement to go out and get moving. The needles and yarn can feel like a safe extension of our selves into the world, and I'm sure this is the case with other objects used in gentle activities, from fishing rods to paintbrushes.

And movement is just the first part of the knitting equation. Knitting also creates an enriched environment: one of relaxation, meditation, creativity and purpose. In a similar way to hypnosis and virtual reality, the calm but concentrated focus required for knitting – alongside strong visual and tactile stimulation – can distract, dissociating the pain in the brain in the short term, and diminishing it in the long. A 2019 study at the Mayo Clinic in Florida monitored people's brainwaves using electroencephalography (EEG) while they knitted and detected a pattern of theta waves, a type of brainwave associated with meditation.[6] The process of creating something gives the individual a sense of purpose, adaptability and control. Creativity helps train the brain to adapt and adjust to an uncertain, ever-changing world. Creating something useful also provides reward, esteem, purpose and fun: all things that can so easily be lost when living with constant pain. The last element of the knitting equation is that it can be immensely social. In the knitting groups she has set up and visited, Corkhill has seen first hand the radical effect socializing has on pain. Isolation and loneliness are well-known pain-worsening stressors, whereas communication, friendship and laughter are all powerful pain-relievers. Finally, knitting is a cheap and extremely portable intervention. For long-lasting effects on reducing pain, the process needs to be repeated again and again, ideally a few times a week, for the brain to learn and develop, and to break out of the bad habit of long-term pain. As knitting tends to involve working towards a completed project, it really lends itself to this. As patients knit, they really are taking part in a constructive, automatic, positive habit. They are knitting new neural pathways in their brain, stitch by stitch. The words of one knitter, interviewed in one of Corkhill's studies, get to the core of why knitting has been successful for them: 'I am convinced that knitting has somehow reset my brain. The repetitive, meditative, and creative aspects are what has helped me back to a more fulfilling life.'[7]

I am of course not saying that knitting is the solution to persistent pain. It helps some people, but not others. It also needs more research to consolidate a strong evidence-base, and at the time of my interview with Betsan Corkhill much of this new research had been frustratingly halted by the Covid-19 pandemic. But what I love about knitting is what it represents – it helps people to actively tackle many aspects of pain at the same time. My own knitting skills leave much to be desired and my attempts inevitably end up as a shapeless, multicoloured mass, something that would make the Gordian knot look like a shoelace. Picture persistent pain as one of these complex, seemingly intractable knots I unintentionally create. It's made up of various intermingling strands – stress, relationships, anxiety, other health conditions, past experiences, to name just a few. To truly tackle persistent pain – to create an environment of safety in our brain and body – we need to take hold of a number of these different strands and slowly unravel the knot.

Pain isn't found in the body, but it also isn't just found in the mind: pain is in the person. To treat pain, we need to treat the whole human. Recovery is about changing the meaning of pain; it is about recovering identity and personhood. This might seem daunting, but it means we have many opportunities, many different points of entry. Pain is a complex system that's influenced by, interwoven with, all aspects of our life and our world. But our approach to complex systems doesn't need to be complex. Seemingly simple changes can have exponentially powerful effects on our experience of pain. These changes may seem boring, or may not seem 'medical' or 'scientific' enough – such as sleep, exercise and socializing – but their knock-on effects can be revolutionary. Simple things can have a big impact, often with unintended positive consequences. For example, before the Covid-19 pandemic, handwashing wasn't taken *that* seriously by the general population. In all honesty, even the average healthcare worker in your average hospital didn't take it too seriously (with the exception of surgical-theatre staff). But the cultural change in handwashing in 2020 not only helped slow the spread of Covid-19, but also reduced the transmission of many other viral and bacterial infections across the world. Hong Kong's tough response to the new coronavirus also ended their flu season two months early, and cases of *C. difficile* infection – a hospital-acquired bowel infection that causes severe diarrhoea – dropped by 70 per cent in Spanish hospitals.[8] And while small changes can have big effects, scientific studies probably underestimate the power of multiple positive changes in someone's life, even when the effects of each intervention on their own are moderate. Trish Greenhalgh, Professor of Primary Care Health Sciences at Oxford University, put this wonderfully in a Twitter post: 'Literature on complex systems highlights a more organic model of causality with multiple interacting influences. Ask not "What is the effect size; is it statistically significant, controlling for other variables?" but "Does this intervention *contribute* to a better outcome?" Using the logic of complex systems, multiple interventions might each contribute to an overall beneficial effect, even though none of these interventions individually would have a statistically significant impact on any predefined variable.'[9]

Before taking a look at the most promising tools for tackling persistent pain, let's briefly look back into the heart of pain, to see what we are healing, and how. At the deepest, neurological layer of the onion, we are getting the brain – often with help from the body – to change itself. Persistent pain is caused by our changeable, neuroplastic brain becoming more efficient at creating pain, even when there is no longer any danger. We can reverse these changes by slowly but steadily retraining our overprotective brain to be less protective. One level up, we make these changes in our brain both indirectly, by reducing stress and inflammation, and directly, by using techniques to retrain the pain system. In short, treatments that work are treatments that make the brain feel safe in its body. I like to think of it as soothing the brain.

What follows is a selection of persistent-pain treatments. It is not an exhaustive list – this is not a self-help book – but between them these methods encapsulate some key takeaways. I've placed them in three broad categories:

Alteration: Altering our brain's context (through body, mind and environment) for it to feel safe.
Visualization: Weakening pain by stealing back your brain.
Education: Knowledge is power.

These interventions – ranging from the seemingly mundane to the hard-to-believe – might appear unrelated on the surface, but they all follow the principle of retraining our brain to feel safe in its body.

Alteration

First, I cannot overstate the importance of movement. We were made to move. Sedentary beings are vulnerable beings, and immobilization increases our sense of fear. Prolonged immobility is a multi-system disease: it amplifies pain, degrades muscle and bone structure, lowers mood and wreaks havoc on our immune and cardiovascular systems.[10,11] It is responsible for years of misery and early deaths; one study suggested that sedentary behaviour was related to 70,000 deaths a year in the UK.[12] All evidence shows that the risk of immobility is greater than the risk of activity. And the good news is that movement is low-cost, safe and effective at reducing pain, with the side effect of improving physical and mental health.[13] Movement and exercise normalizes the balance of almost all body systems (known as homeostasis) more than any drug or diet. Movement strengthens the body, lubricates joints, removes waste products in tissues and opens the brain's drug cabinet of natural pain relievers. Movement can directly reduce pain in the short term,[14] it is anti-inflammatory,[15] improves sleep, nourishes mental health and – contrary to popular belief that movement is tiring – reduces fatigue and increases energy levels.[16] Most crucially, movement can train the brain out of persistent pain. If we move in a way that makes our body feel strong and healthy, our brain is given regular signals suggesting that our body is strong, healthy – and safe. Slowly, but surely, the brain relaxes and becomes less vigilant; this is healing our brain from the bottom up. In the words of one of my favourite age-old axioms: motion is lotion.

Now, this does seem counter-intuitive, and a common (and understandable) objection to movement for those already in constant, persistent pain is this: it hurts. That's true, and it's certainly never easy at the start. But it's so important to remember the central truth about persistent pain in most instances: even though it hurts, your body is not being damaged. The brain has become overprotective and we feel pain long after the injury has healed, so we can be sore yet safe. Understanding that it is safe to move is just as important as moving. This is why great results are found when exercise interventions are combined with pain education.[17] I'm not suggesting, however, that to cure persistent pain you have to bench-press your bodyweight or sign up to an Ironman triathlon. Movement needs to start off very, very gently and gradually be built up with baby steps, slowly increasing strength, mobility and coordination. 'Graded exposure' is a common-sense approach in which you gradually make movements that give the tissues a workout, then over time the body tissues strengthen and the brain interprets these movements as safe. Alongside this, regularly performing painful movements at low intensities sends messages to the brain that movement is safe, leading to a disassociation between the movement and the sense of threat. This provides the brain with positive information about what is going on in the body. It's not too dissimilar from exposure therapy for phobias: getting you to first see the spider, learn that it is not dangerous, then incrementally be exposed to it so that within a few weeks you are holding the spider in your hands and using it to scare your friends.

Gently pushing the limits of movement is also so much easier when it is fun, creative and meaningful, whether it's walking in a beautiful forest, playing a team sport or de-stressing in a Pilates, yoga or tai chi class. One of my personal favourites – for those lucky enough to have access to a pool – is anything 'aqua': aqua-jogging, water aerobics or swimming. I am thoroughly biased, having spent most of my youth swimming, but there's a lot of evidence for its health benefits[18] and the supportive buoyancy of the water is the perfect environment for movement with a feeling of safety.

Another cheap and effective way of making the brain feel safe in its body involves something we do very often. In fact, if we stopped doing it we'd be dead in a matter of minutes. Breathing is obviously crucial for survival: it's how we get our oxygen and expel carbon dioxide, the waste product of our body's metabolism. But the *way* we breathe was something I never came across at medical school. I always assumed that breathing exercises might be helpful for the very anxious, but that otherwise they weren't really 'proper' medicine. I didn't realize how wrong I was until breathing exercises found me. My first year of life as a junior doctor involved long periods of immense stress. On one particular night – the fourth in a string of thirteen-hour night shifts – I had assessed and treated hundreds of patients and, sadly, had to certify the deaths of three patients, one of whom was on the ward I usually worked on and I had got to know really well over the previous few weeks. As things quietened down at around 4 a.m., I slumped into a chair in one of the ward offices, fired up the computer and began an intense search for any advice for reducing my stress, and quickly. My scepticism (or ignorance) was quickly extinguished by the overwhelming evidence for breathing exercises.

Healthy breathing is slow and deep, and properly engages the diaphragm, that tent-like muscle at the base of our lungs. But in our modern world of persistent, low-grade stress, most of us breathe too shallowly and quickly. In the primaeval fight-or-flight response, our sympathetic nervous system kicks in and tries to make our breathing quick and shallow in an attempt to increase oxygen levels. This is all good in the short term, but in the long it messes with the balance of oxygen and carbon dioxide and wastes energy through an inefficient use of chest muscles. The solution to this is simple: breathing deeply and slowly. In short, breathing deeply stimulates the vagus nerve and activates the parasympathetic nervous system, which helps us both rest and digest. It's the relaxation response. To breathe well, there are many variations on a theme, but the core of it is simple. First, find a relaxing place to sit or lie down, then close your eyes and your mouth. Take a slow, deep breath in through your nose for five seconds, letting your belly expand outwards, pause for one second, then let out a slightly longer, seven-second breath. On inhaling, only the belly should be moving. This technique can be tricky at first but is easily mastered. The key is to do it regularly – perhaps a set of ten to fifteen breaths three to five times a day. In numerous studies it has been shown to reduce stress, inflammation and persistent pain.[19,20]

There are many other tricks to get the body to relax the brain. Seeking relaxing, novel sensations, such as from a hot bath, a relaxing massage or the vibrations of an electrical stimulation machine (like a TENS machine) can help. These interventions are not necessarily changing the body, as the tissues have already healed. Instead they are sending comforting and distracting sensory inputs to the brain. However, it's very important not to rely solely on 'passive' treatments – where you lie down and receive treatment from a therapist or machine. If used as the only treatment for pain, they discourage self-growth and self-management. But they can be an extra means of relaxing and letting the brain feel safe in its body.

What we do with our body can help heal persistent pain, but equally important is what we put into it. Chapter 10 supplied overwhelming evidence that smoking, obesity and alcohol excess significantly increase inflammation and worsen persistent pain. Making life changes that involve eating healthily and cutting down on nicotine, alcohol and caffeine are not as easy as taking a pill and require time, effort and support. These interventions are neither 'quick fix' nor romantic, but they really work.

Getting the body to rest and relax is clearly crucial to calming an overprotective brain and stressed-out body. The greatest form of rest – sleep – often plays a central role in people's persistent pain, yet it's often overlooked. It might seem obvious that someone in constant pain will find it hard to sleep, but a lack of sleep itself worsens pain. In fact, there's evidence that – counter-intuitively – insomnia causes pain more than pain causes insomnia.[21] This relationship is incredibly important. For some people, proactively improving a night's rest may have revolutionary effects on their pain. This can start with being disciplined with sleep hygiene, which involves cultivating good habits and optimizing the sleeping environment. For example: make sure the bedroom is dark and quiet; keep a regular nightly routine that involves winding down – keeping screens outside of the bedroom – and avoid evening caffeine; have a fixed wake-up time. If changes like these make little difference, cognitive behavioural therapy for insomnia (CBT-I) improves sleep quality and insomnia symptoms in 70–80 per cent of patients,[22] and it also improves insomnia-related persistent pain.[23] Looking after our body is a surprisingly effective way of communicating with our brain – letting it know that things are OK and it does not have to be on hyper-alert for danger. Safe movement (with relaxing, enjoyable rest), safe breathing, safe sleeping and healthy lifestyle choices are interventions that are seemingly very different to one another on the surface. But the principle is the same. They all get to the heart of pain through two simple mechanisms: reducing stress and nurturing a sense of safety.

Another way of talking to our pain system and politely letting it know that it can have a rest is through our mind. As we saw in Chapter 5, while we clearly can't just think pain away by putting on a smile, our mind can have a huge influence on the experience of pain. Slowly but surely reframing our mindset from one of fear and stress to one of confidence and hope really does improve the experience and intensity of pain. Nurturing an outlook of mental flexibility, a positive openness to change and an acceptance of what has happened is crucial. Acceptance seems counter-intuitive, because it is. But instead of giving up and giving in, acceptance is about reaching a point of understanding of what pain is and coming to terms with the position you are in. Indignantly fighting an unknown enemy only causes more stress and worsens everything. Pain is our protector, our guardian angel. Yes, it's dreadful when it is overprotective, but it will not be calmed down by denial or anger: we need to speak its language and tell pain – from our body and mind – that we are safe. Reaching this point may involve talking therapies with a knowledgeable and skilful practitioner, such as cognitive behavioural therapy (CBT), mindfulness-based stress reduction (MBSR) or acceptance and commitment therapy (ACT). New psychotherapies, such as emotional awareness and expression therapy (EAET) – which helps people to process stressful life events – also show promise. The bottom line is that a talking therapy needs to affirm modern pain science, reduce fear and increase confidence. In cases of past trauma or formal mental health diagnoses, having these addressed by an understanding psychologist or psychiatrist can be a powerful pain reliever. And, most importantly, being creative – in doing things you enjoy and that provide meaning – is key to growing on the journey.

There are plenty of avenues to reduce stress and foster a sense of safety through our body and our mind. But no one is an island. We are deeply linked to our environment, through our relationships, our financial situation, our perceived position in society and even the architecture of our home and workplaces. Chapter 9 showed how social context directly affects pain in numerous, often unexpected ways. Doctors and other healthcare professionals are frequently unequipped to deal with social factors, and can often give the impression that if the cause isn't medical, it isn't *really* that important. They couldn't be more wrong. Healing from pain may require directly addressing external life stressors, whether at home, at work or in other relationships. This is easier said than done, but directly dealing with these may directly improve persistent pain. When it comes to eliminating life stressors, it's worth looking for the lowest-hanging fruit first – the stressors that are easiest to change. In his excellent book *Why Zebras Don't Get Ulcers*, Robert Sapolsky suggests applying the 80/20 rule to stress management: '80 per cent of the stress reduction is accomplished with the first 20 per cent of effort.'[24]

The person in pain should never go through pain alone. Connecting with others by joining groups (and not just chronic-pain groups) is a powerful pain reliever. Seeking to engage in acts of kindness and gratitude to others as well as yourself is vital, for both people in pain and those supporting them. But the social element of pain also starkly shows that, while self-management is crucial for the persistent-pain journey, it is not just the responsibility of the person in pain to reduce their stress. While many people living with persistent pain have the ability and resources to make changes to their lifestyle, if you told a homeless person in the grip of alcohol addiction or an unemployed single parent with a history of trauma to simply 'find ways of reducing stress', you would justifiably be given an earful or shown the door. We all have a responsibility – at all levels of society – to support and empower people.

The brain's decision to make pain is largely outside of our conscious control. We don't have direct access to the corridors of the brain's Ministry of Defence. If it thinks we are in danger, even if it is basing its decision on false information, it will let us know. Altering the context in which the brain lives – whether physical, mental or social – may seem mundane, but it often works at soothing pain, as it nurtures a sense of safety, sending streams of 'good news' to the brain's pain system. Imagine a set of scales in the brain: on one side of the brain it places the evidence for safety and security, on the other it places all the available evidence of stress and threat. Pain is caused by a tipping of the scales to the latter side. The more you increase the brain's sense of safety (e.g. by movement) or you reduce stress (whether mental, social or inflammatory stressors), the more the brain's pain system calms down.

Visualization

While changing the way the brain feels about its body and its environment is an indirect way of calming a pained brain, there is a small but growing number of clinicians and researchers who believe that we can directly access, and ultimately change, our pain system. If we could directly rewire our brain – reversing the neuroplastic changes that bring about pain – it could revolutionize the treatment of persistent pain. One of the most fascinating 'neuroplasticians' is Dr Michael Moskowitz, a pain specialist based in Sausalito, California. He argues that harnessing neuroplasticity can diminish, and even cure, the most stubborn, intractable persistent pain. Moskowitz is not a New Age guru at the fringes of medicine; he started his medical career as a psychiatrist and ended up specializing in pain medicine, becoming one of America's top pain physicians. Moskowitz has successfully treated people who have tried every medical and surgical avenue and failed, the first of whom was himself. In his interview with Norman Doidge, in Doidge's 2016 book *The Brain's Way of Healing*, Moskowitz describes how a life-changing accident was the stimulus for his path to healing and, ultimately, helping others.[25] In 1994, the forty-four-year-old doctor was having a well-earned holiday with his daughters. They went 'tubing', a variant of waterskiing in which the skis are replaced with an inflated inner tube on which you lie and cling on to for dear life, completely at the mercy of the motor-boat driver. As he was skimming over the water's surface, Moskowitz was suddenly flipped off the tube, hitting the water at 60 kilometres per hour with his head tilted back. He spent the next decade in tormenting, agonizing persistent neck pain. None of the treatments he tried – from strong opioids to physical therapy – came even close to touching it. It was a classic case of persistent pain: the pain became gradually worse over time and slowly spread to both sides of his neck, then to his upper back and shoulder blades. Each time he felt a twinge of severe pain, his brain was becoming better at recognizing it and firing up the same pain circuits. It was neuroplasticity gone wrong.

In 2007, after thirteen miserable years of pain, Moskowitz decided to do everything in his power to understand persistent pain in a bid to defeat it (including reading fifteen thousand pages of neuroscience) and came to a simple but profound conclusion. If neuroplasticity had got him into this mess, he could use neuroplasticity to effectively 'learn' his way out of it. In persistent pain, pain circuits spread and effectively steal neural real estate from many areas of the brain, affecting sensation, emotional regulation and cognitive function. Moskowitz's plan was to steal back these areas by overwhelming them with non-painful neural activity, disconnecting pain networks. His weapon of choice: vision. A significant part of the brain is given over to processing vision, and visual input plays an important role in the experience of pain. Moskowitz wondered whether intense and repetitive visualization would be enough to rewire the painful brain.

He first drew three pictures of his brain: one in short-term pain, with various areas of the brain activated, one in persistent pain, with those areas expanded, and a third picture of his brain in no pain at all. Whenever he had a twinge of pain he would close his eyes, visualize the picture of his brain in persistent pain and imagine the activated, pain-creating areas shrinking. This was not simply wishing pain away; in the first three weeks he didn't notice much of a change at all. It took a month for him to automatically apply the technique whenever pain reared its head, then something remarkable happened. By six weeks, the pain across his back had disappeared and, over the course of a year, his neck pain began to diminish – then it completely vanished. His brain had formed a new body-image map – one that wasn't dominated by pain. Moskowitz understandably went on to zealously share this newfound knowledge with his patients, and saw astonishing, seemingly miraculous results. He has dedicated the rest of his career to using neuroplasticity to rewire the brain out of pain. It's not an easy process, particularly at the start. People have to be motivated and have to relentlessly apply the technique, even if there are no rewards in the first weeks or months. It's like learning a new language. But those who keep going often reap rewards of intense and – importantly – long-lasting relief.

When I first read Dr Moskowitz's remarkable story I was slightly sceptical, but then I thought back to when I was completely cured of irritable bowel syndrome through self-hypnosis. Perhaps it was the visualization aspect of hypnosis – imagining my bowels changing from rocky rapids to the languid Oxfordshire Thames – that had the most powerful effect. Not to mention the power of visual illusion created by the mirror box we came across in Chapter 11, which can retrain a brain out of phantom limb pain and even other persistent pain conditions when used in a graded motor imagery programme. Most excitingly, perhaps the emerging technology of virtual reality (VR) will make visualization not only more powerful but much more accessible. A wonderful 2008 study found that in individuals with chronic hand pain, putting a magnifying glass over the hand worsens their pain, and minimizing the size of the limb relieves it.[26] When the brain sees a smaller hand, it assumes that there is less damage. A 2018 study went even further, showing that knee osteoarthritis pain is significantly reduced using VR illusions.[27] A team from the University of South Australia asked patients with osteoarthritis to don a VR headset that showed them a live video of their knee. The VR's software, however, could make the knee appear either smaller or larger. While the participants looked down at their knee, scientists held the calf muscle below the knee and gently either pushed towards the knee or pulled towards the foot. This combination of sight and touch created a 'visuotactile illusion' of the knee either contracting or expanding. This multisensory illusion worked: repeated illusions resulted in a 40 per cent reduction in pain. Moskowitz also found that not everyone responded to visualization alone; some were hugely helped by the addition of touch, sound and vibration – elements introduced to Moskowitz by American physician and osteopath Marla Golden – which added a pleasurable sensory barrage to help unweave stubborn pain circuits.

The idea of being able to steal back your brain, to reclaim the territory taken over by persistent pain, is tantalizing. The prospect of directly retraining the brain to relieve pain looks promising and is in keeping with a modern understanding of pain. Neuroplasticity research and clinical practice is in its infancy, though, and we need more evidence-based scrutiny before shouting its praises from the rooftops. But watch this space . . .

Education

I believe this to be the most important therapy. We cannot rewire a system we do not understand. Understanding how pain works – in a way that is accessible for anyone in pain – is necessary for living with, and relieving, pain. It is a map of the pathway to healing. For many people living with persistent pain, pain education can be more powerful than drugs and can lay the foundations for actively seeking evidence-based lifestyle changes and informed treatment choices. Pain education is needed because most of us (including many health professionals) believe the outdated and outright wrong view that pain is produced in the tissues and detected by the brain. Changing one's framework to one fed by modern pain science – that pain is an output of the brain, our protector and defender, and not an informant of tissue damage – requires a complete conceptual shift: a pain revolution.

Communicating the central truths of pain in a clear, memorable and evidence-based way is no easy task, but it has been done, and with promising clinical effects. One of the pioneering works in modern pain education is *Explain Pain*, a book and educational course created by the Australian pain experts (and expert pain explainers) David Butler and Lorimer Moseley.[28] Its playful use of story and metaphor, and the development of new tools and concepts, is often ingenious. Take, for example, 'The Protectometer', a 'danger meter' people can use to identify contributors to, and relievers of, their pain, which the authors have respectively termed 'Danger In Me' (DIMs) and 'Safety In Me' (SIMs). This encourages individuals to seek out things, people and experiences that their brain sees as credible evidence of safety. Most importantly, there is good evidence that it works. In one study carried out soon after *Explain Pain* was first published, participants experiencing persistent back pain were randomly assigned to either an *Explain Pain* group, in which they were taught the modern understanding of pain, or a traditional 'Back School' group, in which they were taught about spinal anatomy, physiology and ergonomics.[29] Those in the *Explain Pain* group had immediate pain relief but those in Back School actually experienced more pain. Modern pain science had given people the confidence that pain does not always mean tissue damage, whereas the traditional group had just learned the names of various parts of their back that could be damaged.

In recent years there have been a number of randomized controlled trials assessing the efficacy of *Explain Pain.* They have been of varying quality and come to slightly different conclusions, but a picture is emerging: not only does *Explain Pain* improve pain knowledge, decrease activity-related fear and increase active engagement in rehabilitation, but it actually reduces pain.[30,31] Importantly, these effects are seen even in the long term, with one study concluding, after a twelve-month follow-up, 'an improvement in pain biology knowledge was significantly associated with a reduction in pain intensity'.[32] And this is the result from pain education alone; if the effects were the foundation on which the person in pain sought other treatments mentioned in this chapter, the likelihood of pain reduction is even higher.

Explain Pain is one of a number of pain education platforms, and one of a few with a good evidence-base. It would be wonderful if people in persistent pain could also use an easy-to-access and evidence-based app to help them on their journey. One app that caught my eye is *Curable,* set up by three individuals who have each recovered from about a decade of persistent pain. They have great pain scientists on their advisory board, and the platform provides modern pain-science education and a virtual coach to aid active self-management. Of seven thousand *Curable* users questioned, 68 per cent experienced real pain relief after thirty days. At the time of writing, the research and methods of this study aren't fully published, so we can't start counting our chickens yet, but I hope that this and similar programmes can be shown to aid people in their self-management of pain, and enable them to diminish their pain or live a fulfilled life alongside it.

Education is not just for people in pain. It is utterly astonishing that most doctors do not have a real grasp of the nature of pain. I was certainly one of them. And while many do, the system is not set up to understand it. Healthcare in the West is designed around individual diseases in neat and organized silos. Patients are seen by different specialists for different body systems and – in the greatest divide of all – go to one hospital for their body and another for their mind. Medical schools focus on the biomedical model of pain, but its powerful cognitive, psychological and social influences get only a token, passing mention at best. While medical students are deeply concerned with patients in pain, they find chronic pain the most difficult aspect of pain to deal with.[33] They are taught about pain mechanisms – and doctors are usually endowed with a great knowledge of how to manage acute pain in a hospital setting – but not how they are relevant to persistent pain. Most doctors like to diagnose a patient with an interesting – but understandable – disease with an identifiable cause, then prescribe or deliver a treatment that works and see the patient get better. Case closed. But persistent pain is messy, complex and human.

If healthcare providers don't accept the reality of pain, they won't see the possibilities of treating it. Worse, we can unwittingly heighten people's pain by reinforcing the fearful belief that some unidentified tissue in their body has been damaged. The artificial boundaries in healthcare, the power of the pharmaceutical industry and an incorrect understanding of pain has got us into a situation where most people expect their persistent pain to be cured by a pill or their tissues fixed by surgery. For most, this doesn't work and creates helplessness and hopelessness. But there is a better way. If we educate others and ourselves about the true nature of pain, we can empower, encourage and heal. Like most scientific revolutions, changing the system will be a slow process. It feels like trying to change the course of a river. But we will do it, one transformed life at a time.

I spoke to one pain specialist who is already pushing for change. Dr Deepak Ravindran is a consultant in anaesthesia and pain medicine at the Royal Berkshire Hospital in Reading, England. He has over twenty years' experience in pain management, an expert knowledge of pain-relieving drugs and practical skills in various injections and nerve blocks. He is a child of the biomedical school of pain but, converted by the evidence, is joining the pain revolution: 'Whole generations of people, including doctors, have been brought up to think that pain is a sign of damage. Most people believe that scans will always find the right cause of the damage, and all that needs to be done is for this tissue to be blocked, cut, taken out or numbed. When I was training, opioids were sold to us as effective and non-addictive, but around 2014 and 2015 it was becoming clear that this was not the whole story. When I look at the evidence and the experience from my clinic, drug therapy is only about 30 per cent effective, 30 per cent of the time, in about 30 per cent of my chronic-pain patients. Around the same time, I was also reading convincing evidence that a number of types of surgery that aim to relieve pain – from shoulder arthroscopy to facet joint injections – often perform no better than a placebo . . . This has really humbled me as a pain physician. We need to move from a purely biomedical model, of pain being a symptom of damage, to a wider understanding of pain as a protective mechanism.' Dr Ravindran hasn't thrown away his drug cabinet or stopped practising interventional techniques, however. 'We need to separate nociception – danger detection – and pain. If there's evidence that pain is caused primarily by nociception coming from a damaged body part, then traditional methods may be appropriate. Otherwise, we need a broader view of pain.'

Dr Ravindran's discovery of the true nature of pain – and what that means for pain relief – has led him to become a passionate pain educator. In his 2021 book, *The Pain-Free Mindset*, Dr Ravindran sets out to explain pain to people living with persistent pain and, in light of our modern understanding of pain as a protector, lays out seven areas through which we can find relief. These neatly fit the acronym 'MINDSET'. M and I stand for 'medications' and 'interventions', the traditional pain-relieving methods used by the medical profession. Dr Ravindran is clear that, while being ineffective in the majority of cases of persistent pain, they can be of use to some people and can complement other therapies. He is also adamant that these interventions are useless if given passively and if the person in pain does not understand their pain and take control of their health. This is why the next letter in the acronym, N, stands for 'neuroscience education'. The next three letters – D for diet, S for sleep and E for exercise – explore evidence-based ways in which we can reduce pain by reducing inflammation and making the brain feel safe in its body. T stands for 'therapies of mind and body', ranging from cognitive therapies to therapies for an area often overlooked by the traditional model of pain – trauma. Dr Ravindran carried out an audit in his practice that found that 40 per cent of his patients had experienced significant adversity in childhood. For many of them, treatments that don't address the underlying brain changes caused by trauma are merely scratching the surface.

Dr Ravindran's work chimes with what we have covered in this final chapter. He understands that pain is personal and that approaches to treatment have to be personalized. Pain wants to protect the whole human, and to reduce pain and make the brain feel safe we need to address all aspects of human experience. This is nicely summarized by the Stanford pain specialist Sean C. Mackey: 'Chronic pain is not all about the body, and it's not all about the brain – it's everything. Target everything. Take back your life.'[34]

There are no quick fixes. The path up the mountain of healing is narrow, winding and often steep. It is a journey with storms and false summits. But the evidence shows that it is a journey that, if followed with persistence and hope, is worth it. Let's be kind to the brain and the body, and educate to empower. A modern understanding of pain teaches us not to see people as their pain, nor to see people as a bundle of receptors and nerves, but to see people as humans. To understand pain is to understand ourselves. I hope that this book has whetted your appetite. Please read further, spread the truth and, most importantly, have hope.

Glossary

Acceptance and commitment therapy (ACT)
A psychological intervention that, at its core, involves the individual learning to accept the distressing situation they are in without trying to resist or fight it. Once a level of acceptance has been achieved, ACT helps people to reframe the way they think about pain, as well as to evaluate their life goals and values. In a way, ACT doesn't directly try to reduce a person's symptoms; a reduction in pain is a by-product.

Acute pain
Short-term pain, usually associated with a stimulus, whether touching a boiling kettle or stepping on Lego left on the carpet. If the stimulus causes injury, acute pain resolves once the injury has healed.

Allodynia
Pain caused by a stimulus that does not usually cause pain, and often caused by pre-existing damage or inflammation (see 'Inflammation'). Think of putting a shirt onto a sunburned back.

Amygdala
An almond-shaped structure ('amygdala' comes from the Latin for almond) located deep within the temporal lobe of the brain. Humans have two, one in each hemisphere of the brain. The amygdala plays a key role in processing fear and threat stimuli, and helps trigger the fight-or-flight response.

Anandamide
A molecule made by our bodies that brings about pain relief and pleasure by activating cannabinoid receptors in the brain (see 'Cannabinoid'). *Ananda* is the Sanskrit word for bliss.

Anhedonia
A reduced ability to desire or enjoy pleasurable stimuli, such as eating or sex. This is a common component of depressive disorders.

Anterior cingulate cortex (ACC)
A boomerang-shaped area of the brain that, roughly speaking, sits between the 'emotional' and 'cognitive' areas. The ACC seeks not to understand the intensity or location of pain, but instead what it *means*. Among many other functions, it evaluates and integrates the physical, emotional and social elements of pain. When someone hurts our feelings, or we feel the pain of exclusion, we have our ACC to thank.

Axon
The branch of a nerve that travels away from the cell body. These are in essence the main electrical wires of our brain and nervous system.

Basal ganglia
A group of structures deep within the brain that are responsible for a variety of functions, from controlling movement to emotional function.

Bayes' Law
$P(A|B) = (P(A)\ P(B|A))/(P(B))$. If that hasn't cleared it up, Bayes' Law is a formula that works out the probability of an outcome occurring based on a previous outcome occurring. It enables us to update future probabilities based on new information. It was devised by Thomas Bayes, an eighteenth-century English Presbyterian minister. He didn't live to see his discovery published.

Bio-psycho-social model
First proposed by the American physician George Engel, this theory illustrates that disease (and health, for that matter) is not purely biological but is profoundly influenced by psychological and social factors.

C-fibre
A type of nerve fibre that lacks a fatty, insulating sheath of myelin (see 'Myelin'), so it transmits nerve impulses at a slower speed than other nerves.

Cannabinoid

Compounds found in the cannabis plant. There are over one hundred of these, and many display a wide variety of effects in humans, from euphoria to increased appetite. While cannabinoids can be extracted from the plant, they can also be made synthetically. Our body even produces its own cannabinoids, known as endocannabinoids.

Capsaicin

The active compound in chilli peppers. These activate the same receptors in our mouth and on our skin that detect hot stimuli. Capsaicin tricks our brain into believing that our body is being burned, even though there's no increase in temperature.

Cauda equina syndrome

Right at the lower end of the spinal cord is a bundle of nerves that resembles a horse's tail – the Latin for which is 'cauda equina'. If this structure is damaged it can cause severe back pain, numbness in the saddle area, sciatica-like pain, incontinence and sexual dysfunction. Such damage is considered an emergency and is treated with surgical decompression.

Central nervous system (CNS)

The brain and the spinal cord.

Central sensitization

An increased responsiveness to danger signals in the brain and spinal cord. This results in an increased sensitivity to danger signals, as well as feeling pain from stimuli that don't usually cause pain. It's likely that central sensitization is a main mechanism in the transition of acute to chronic pain.

Chronic pain

Long-term pain. Definitions vary: some say that it's pain that lasts more than three months; others that it is pain that persists after an injury has healed. Also known as 'persistent pain', a term I prefer and use more frequently (see 'Persistent pain').

Cognitive behavioural therapy (CBT)
This is a psychological therapy that aims to equip people with the skills to change negative thoughts and behaviours.

Cognitive functional therapy (CFT)
A mind–body approach to pain that helps individuals to reframe their mindset and 'pain beliefs' through pain education, movement and changes to their lifestyle.

Complex regional pain syndrome (CRPS)
As its name suggests, a complex – and poorly understood – condition of severe pain that is usually confined to one limb. Research strongly suggests that this is caused by inflammation (see 'Inflammation') and central sensitization (see 'Central sensitization').

Congenital analgesia
Also known as congenital insensitivity to pain. Any of the rare genetic conditions that result in an individual being unable to perceive pain. One cause of this is the failure of nerves to send danger signals to the brain.

Covid-19
Coronavirus disease 2019, abbreviated to Covid-19, is an infectious disease caused by the 'severe acute respiratory-distress syndrome coronavirus 2' (SARS-CoV-2). The number 2 is used to differentiate it from SARS, another coronavirus known for its deadly outbreak at the turn of the twenty-first century. It brought the world to a halt and caused millions of deaths at the turn of the 2020s.

Cricket
As you may have gathered, I don't really understand the game. A cricketing friend once told me an 'oldie but goodie' definition of the rules: 'You have two sides, one out in the field and one in. Each player that's in the side that's in goes out, and when they're out they come in and the next player goes in until they're out. When they are all out the side that's out comes in and the side that's been in goes out and tries to get those coming in out. Sometimes you get players still in and not

out. The game ends when both sides have been in and out including the not-outs. Got it? Good.'

Damage-associated molecular patterns (DAMPs)
Molecules in our body that are released after tissue damage and which can be recognized by our immune cells, activating the immune system (see 'Immune system'). These include proteins released from damaged or dying cells.

Deep brain stimulation (DBS)
A treatment involving neurosurgically implanted electrodes that release electrical impulses in specific brain regions. These 'brain pacemakers' are intended to regulate abnormal impulses.

Default mode network
A group of brain regions that appear to show more activation when we are at rest and less activation when we are engaged in an activity. They tend to be involved in recalling memories, daydreaming and planning for the future.

Dendrite
A branch-like part of a nerve that receives inputs from other nerve cells and passes this information to the cell body. Usually, multiple dendrites lead into one cell body; indeed, the word derives from the Greek word for tree: *dendron*.

Dopamine
A chemical messenger that enables many different responses in the brain, from movement to motivation. It is commonly mistaken for being involved in generating the feeling of pleasure from a reward, whereas it plays more of a role in motivating us to seek the reward first.

Electroencephalography (EEG)
A non-invasive way of recording electrical brain activity. Electrodes placed across the scalp measure changes in electrical current across the brain. It is best known for its use in the diagnosis of epilepsy, but it has many other clinical and research functions.

Empathy
Definitions and variations differ, and there are subtypes of empathy too, but by and large it is the ability to feel and understand what someone else is experiencing.

Emotional awareness and expression therapy (EAET)
An emotion- and trauma-focused talking therapy that aims to help people understand the importance of emotions in affecting brain pathways that amplify pain, particularly in relation to past trauma or current conflicts. The therapy empowers the individual to express both positive and negative emotions and to adapt to be able to work through problems.

Endocrine system
A system of hormones and hormone-releasing glands throughout the body. Well-known endocrine hormones include cortisol and the sex hormones oestrogen and testosterone.

Endogenous
Originating from within the body. For example, we produce endogenous opioids (see 'Opioid') such as endorphins.

Entorhinal cortex
A small area of the brain's temporal lobe, best known for memory formation. When it comes to pain, the entorhinal cortex plays a role in interpreting potentially dangerous cues in the environment, increasing anxiety and therefore amplifying pain. This is probably to prepare the body to react in the event of a worst possible outcome.

Eye-movement desensitization and reprocessing (EMDR)
A therapy that is often effective for those who have suffered trauma, though we don't fully know why it works. By carrying out a bilateral stimulation task (such as moving one's eyes from side to side) while being asked to recall the trauma, one theory is that the individual can only retrieve limited information about the event and consequently has a reduced emotional response to it. Over time, this desensitizes

the individual to the effects of these traumatic memories by reprocessing them.

FAAH

Fatty acid amide hydrolase, a protein that breaks down anandamide (see 'Anandamide'), reducing the amount of this endocannabinoid (see 'Cannabinoid') in our body.

Fentanyl

A powerful synthetic opioid (see 'Opioid') used for pain relief and analgesia. Being cheap, and often hundreds of times stronger than morphine, fentanyl is now a common recreational drug and one of the most deadly.

Fibromyalgia

A condition characterized by persistent widespread pain, as well as fatigue and memory issues. Frustratingly, we don't yet have a good understanding of its biology, but research is progressing. It's likely that abnormalities in pain processing (see 'Central sensitization') play a role, and recent evidence suggests that some of this could be driven by the immune system (see 'Immune system').

fMRI

Functional magnetic resonance imaging (fMRI) is a form of brain imaging that detects changes in blood flow in the brain. These changes are highly suggestive that the part of the brain requiring increased blood flow is requiring more energy, and is therefore more active. 'Functional' refers to being able to see these changes in brain function, not simply the brain's structure.

Gate-control theory

Scientists Ronald Melzack and Patrick Wall's 1965 theory that the spinal cord contains a 'gate', in the form of intermediary nerve cells, that regulates whether danger signals travelling in from the periphery are passed up to the brain. While we now know that it is an oversimplified theory, it paved the way for modern pain science.

Gout

An inflammatory joint condition, particularly affecting the metatarsophalangeal joint, at the base of the big toe. It's caused by high levels of uric acid in the blood, which form crystals that inflame the joints. While food and alcohol intake can increase the risk of gout – it has been known as 'the disease of kings' due to its prevalence in older men consuming disproportionate quantities of meat and alcohol – diet is only responsible for a minority of cases; genetic predisposition, various medical conditions and certain medications are also risk factors.

GP

A general practitioner (GP) is a doctor who provides a wide range of medical services to patients in the community.

Graded motor imagery

A persistent-pain rehabilitation programme that aims to slowly activate areas of the brain involved with movement without triggering pain's protective alarm. It is a three-step process, first helping individuals to determine whether an image of a hand they are shown is either a left or a right limb, then getting them to imagine moving it, and finally introducing 'mirror therapy', in which the participant sees the reflection of their healthy limb moving, making their painful limb appear to move.

Histamine

A very small compound that packs a big punch. When released by the immune system's mast cells (see 'Immune system' and 'Mast cell'), it causes many of the symptoms of inflammation (see 'Inflammation') and allergy: itching, blood-vessel dilation leading to redness, heat and swelling on the skin, and sometimes a systemic drop in blood pressure. It also causes sneezing and increases nasal secretions.

Homeopathy

A field of alternative medicine built on the theory that 'like cures like', that a tiny, ultra-diluted amount of whatever causes an illness can actually heal it. Clinical trials show that it performs no better than a placebo (see 'Placebo'). Regarded by the vast majority of the medical community to be scientifically implausible.

Hyperalgesia

An enhanced sensitivity to pain-producing stimuli. This can be caused by a number of things, such as nerve damage, inflammation (see 'Inflammation') and even opioid use (see 'Opioid'). Not to be confused with allodynia (see 'Allodynia'), which is pain caused by a stimulus that does not usually cause pain.

Hypnosis

A state in which consciousness is altered to focus attention in one specific direction, which can make someone more responsive to suggestion.

Hypnotherapy

The use of hypnosis to treat a medical or psychiatric condition.

Hypothalamus

A small area of the brain that forms a crucial bridge between our nervous system and endocrine (hormone) system (see 'Endocrine system'). It controls hunger, sleep and temperature regulation, among many other functions.

International Classification of Diseases (ICD)

A list of diseases maintained by the World Health Organization (WHO) to help standardize diagnoses.

Immune system

One of the body's protective systems. The immune system is a beautiful, complex network of cells, molecules and processes that protect the body from foreign threats, whether they be pathogens, cancer cells or inanimate objects that breach the skin.

Inferior frontal gyrus

A part of the brain best known for housing Broca's area, an area crucial to language-processing and the production of speech, though there is increasing evidence that this area also plays a role in the emotional processing of pain.

Inflammation
A coordinated response of the body to harmful threats, from bacteria to broken bones. The role of inflammation is to eliminate the cause of harm and then begin the tissue-healing process.

Insula
An area found deep within the brain, responsible for an astounding variety of functions. These include the feeling of a number of emotions, such as fear and disgust, and the linking of these emotions with the experience of pain.

Internal capsule
One of the brain's motorway junctions. The internal capsule contains bundles of nerve fibres that connect the cortex with other brain structures.

International Association for the Study of Pain (IASP)
The principal global pain society, which brings together clinicians, scientists and policy-makers.

Irritable bowel syndrome (IBS)
A condition characterized by recurrent or persistent abdominal pain and associated with changes to defecation (constipation and/or diarrhoea, as well as changes in stool frequency and appearance). It can be caused or triggered by a number of factors, one of the most significant being psychological stress. Current evidence suggests that IBS results from a dysfunction in the gut–brain axis – the signalling system that connects the gut (and its trillions of microbes) to the brain.

KCNG4
A gene that controls the potassium channel K_v6.4 (see 'K_v6.4').

K_v6.4
A channel (or, strictly speaking, a subunit of a channel) that regulates the flow of potassium across the membrane of nociceptors (see 'Nociceptor').

This enables nerve impulses to carry danger signals from the site of injury to the brain.

Leprosy
A chronic infectious disease caused by the bacterium *Mycobacterium leprae*. Contrary to popular belief, the disease does not cause fingers and limbs to drop off. Instead, once arriving in the human host, the *M. leprae* bacteria seek the cooler climes of human peripheries and populate the nerves of the skin. Damage to these nerves causes the human to lose first their sense of temperature discrimination, then the sensation of light touch, and then pain. Without the alarm of agony, those affected suffer damage to their skin in cuts and burns, the resulting infections (also unfelt) often causing permanent damage to fingers, toes and facial structures.

L-dopa
A molecule that can be broken down to make dopamine (see 'Dopamine'). This is used to increase dopamine in the brain, because dopamine itself cannot be given as a drug, due to its inability to cross the blood–brain barrier.

Lidocaine
Also known as lignocaine, this is a local anaesthetic commonly used in dentistry as well as for minor surgery and uncomfortable medical procedures. It works by temporarily blocking sodium channels in nociceptors (see 'Nociceptor') so that danger signals cannot be initiated.

Mast cell
I like to think of mast cells as the landmines of the immune tissues. Found in most body surfaces that are exposed to the outside world, such as the skin, gut and lungs, when they are activated by pathogens or allergens they release a potent cocktail of histamine (see 'Histamine') and pro-inflammatory molecules, causing swelling, pain and itching.

Mindfulness-based stress reduction (MBSR)
A course comprising meditation, yoga and body awareness, designed to improve emotion-regulation and reduce stress.

Mirror-touch synaesthesia
A condition in which a person, when they see another person being touched, feels the same touch – or what their brain perceives that touch to feel like – on their own body. In some individuals the touch is felt on the same side of the body as the person they see, and in others it is mirrored and on the opposite side. (See 'Synaesthesia'.)

Morphine
A naturally occurring opioid (see 'Opioid'), derived from the opium poppy. It relieves pain by acting on opioid receptors.

MRI
Magnetic resonance imaging (MRI) is a near-miraculous way of imaging the body. One of the building blocks of atoms (and therefore our bodies, and indeed most things in the universe) is the positively charged 'proton'. An MRI scanner is essentially a giant magnet that makes these protons align with its field. While this is happening, the MRI scanner also releases radio pulses, which knock the protons out of alignment. Once the pulse is removed, the protons flip back into alignment, releasing electromagnetic energy that is detected by the scanner, giving us a detailed picture of the different tissues in the body. There are different types of MRI that enable us to view different tissues or functions in the body, such as fMRI (see 'fMRI').

Myelin
A fatty substance that surrounds most nerves, acting as both insulation and as a means of speeding up nerve impulses.

Naloxone
Also known by its brand name, Narcan, this is medication used to block or reverse the effects of opioids (see 'Opioid'). Most commonly used in opioid overdose.

National Institute for Health and Care Excellence (NICE)
A British national body that provides guidance on clinical practice, medicines and health technologies.

Na_v1.7/Na_v1.9
Sodium channels located on the outside of nociceptors (see 'Nociceptor'). These can activate nociceptors and create a nerve impulse – a danger signal – that can travel up to the brain. The channels do this by allowing positively charged sodium to enter the nerve, and the rapid change in electrical charge sets up a nerve impulse. 'Na' is the atomic symbol for sodium, which travels through this channel; 'v' stands for the change in voltage across the nerve's membrane; and 1.7 or 1.9 simply means that it is the seventh or ninth of these channels to be discovered.

Neuron
A nerve cell. Consists of a cell body (an energy-producing powerhouse containing the cell's DNA), dendrite (see 'Dendrite') and axon (see 'Axon').

Neuroplasticity
The remarkable ability of the networks in our brain to adapt and change over time, neuroplasticity was originally thought to be limited to childhood – that intense period of learning – but it is now clear that our brains are fantastically 'plastic' and capable of flexibility throughout our lives.

Neurosignature
There is no specific 'pain pathway' or 'pain centre' in the brain; pain is an output of the brain, created by a pattern of activity simultaneously involving numerous brain regions. This pattern is known as a pain neurosignature. 'Signature' is an appropriate word, as each pattern is unique for every experience of pain.

Neurotransmitter
A chemical substance that transmits a signal from one nerve to another. Dopamine (see 'Dopamine') and serotonin are probably the best known, but there are at least another two hundred others.

NHS
The National Health Service: the publicly funded healthcare system of the United Kingdom, with separately run branches in each of the four constituent countries (England, Wales, Scotland and Northern Ireland). Founded in 1948 to be 'free at the point of use' – this is still largely the case for most services today.

Nocebo effect
When any negative expectations of a medication or treatment regime reduce the positive effects of these treatments.

Nociception
The process of detecting dangerous or damaging stimuli. Nociception is *not* pain! Nociception is neither necessary nor sufficient for the experience of pain.

Nociceptor
Often termed 'danger receptor'. A receptor that detects damage and danger caused by noxious stimuli (see 'Noxious stimulus'), whether thermal, mechanical or chemical. Derived from the Latin *nocere*: to harm.

Noxious stimulus
A dangerous (or potentially dangerous) stimulus to the body. These can be thermal (boiling-hot water), mechanical (a punch) or chemical (corrosive acid).

Nucleus accumbens
A small area of the brain's reward circuitry. It is best known for its role in motivation and in the seeking of pleasurable and rewarding stimuli.

Open-label placebo
An inert pill, which the individual taking it knows is inert. (See 'Placebo'.)

Opioid
A substance that acts on the body's opioid receptors, bringing about pain relief. These can be naturally derived from the opium poppy

(such as in morphine – see 'Morphine' – and codeine, which are also known as opiates), synthetic (such as fentanyl – see 'Fentanyl') or produced by our own bodies (such as endorphins).

Orbitofrontal cortex
A part of the brain that sits just above the eye sockets, in the prefrontal cortex (see 'Prefrontal cortex'). Among other things, it plays roles in decision-making and estimating the relative values of different options.

Pain
The big one! Pain is a horrible feeling that urges us to protect a body part. That's one of my favourite definitions. Any definition should draw from core truths: pain is a protector, not a detector of tissue damage. There is no universally agreed 'final' definition of pain, but the closest thing the scientific community has to a consensus is the IASP's (see 'International Association for the Study of Pain') 2020 definition: 'pain is an unpleasant sensory and emotional experience associated with, or resembling that associated with, actual or potential tissue damage'.

Pain asymbolia
A rare and remarkable condition in which pain is experienced and recognized, but without being aversive or unpleasant.

Paroxysmal extreme pain disorder
A congenital condition of lifelong episodic pain. This is often experienced as rectal pain, but can be felt in any part of the body. Caused by a mutation in the SCN9A gene, which affects the $Na_v1.7$ channel (see '$Na_v1.7$/$Na_v1.9$'), which in turn lowers the threshold for danger signals to be sent to the brain.

Pathogen-associated molecular patterns (PAMPs)
Molecules present in groups of pathogenic microbes that are recognized by the immune system (see 'Immune system').

Pattern recognition receptors (PRRs)
Receptors in the body's immune cells that recognize pathogen-associated molecular patterns (PAMPs – see above).

Periaqueductal grey
A small area of the brainstem that plays an important role in the inhibiting of nociceptive signals coming up from the body. It is one of the last 'gates' that can block danger signals from reaching the brain and being interpreted as pain.

Peripheral nervous system
All of the nerves outside of the central nervous system (brain and spinal cord – see 'Central nervous system').

Peripheral sensitization
Increased sensitivity to nerve stimuli. In the context of pain, this usually happens after tissue damage: think about limping to avoid the pain of putting weight on a sprained ankle.

Persistent pain
Pain that lasts longer than the expected time of healing. This period varies, but in many cases an injury will heal in roughly three months.

Placebo
Anything that appears to be a medical treatment but is in fact inert or inactive, such as a sugar pill.

Placebo effect
More accurately described as 'the expectation effect', this is the brain's response to the context in which a treatment is delivered. If the brain believes that something will reduce pain – whether that's because the patient has read about the treatment beforehand, or because it is being given by a confident-looking clinician – it actually produces pain-relieving chemicals. (See 'Placebo'.)

Placebome
How our genetic make-up can influence our placebo response (see 'Placebo').

Positron emission tomography (PET)
An imaging technique that uses radioactive tracers to detect the levels of oxygenation and glucose consumption in the brain. This shows which areas of the brain are active.

Post-traumatic stress disorder (PTSD)
An anxiety disorder that develops after someone is exposed to traumatic events. It is characterized by flashbacks, nightmares, hypervigilance, negative mood and avoidance.

Predictive processing
A theory that the brain is constantly refining its idea of what is going on in the outside world so it can predict what we are going to experience, with any new or conflicting evidence causing the brain to update its prediction.

Prefrontal cortex
A large part of the brain, just behind the front of the skull. The prefrontal cortex carries out numerous functions, but is best known for 'executive function': decision-making, self-control, short-term memory and control over attention.

Primary erythromelalgia
A congenital condition that causes episodic burning pain, usually felt in the hands and feet. Caused by a mutation in the SCN9A gene, which affects the $Na_v1.7$ channel (see '$Na_v1.7$/$Na_v1.9$') and in turn lowers the threshold for danger signals to be sent to the brain.

Randomized controlled trial
A trial in which groups of similar sizes and demographics are randomly assigned to either: a) have the new drug (or any other kind of treatment intervention) that we want to test or b) have a 'comparator' treatment, whether this be a placebo (see 'Placebo') or 'treatment as usual'. If the study is 'double blind', then not only do the participants not know which treatment they are getting, but the researchers or clinicians do not know which one they are giving, either.

Rheumatoid arthritis
An autoimmune disorder (in which the body's immune system – see 'Immune system' – attacks its own tissues) that mainly affects small joints, particularly the wrists and hands.

SCN9A
The gene that controls the Na_v1.7 channel (see 'Na_v1.7/Na_v1.9').

Somatosensory cortex
An area of the brain, located in a ridge found in the central top area of the brain, responsible for processing touch, balance, temperature and pain. Our brain's maps of the body are stored here.

Synaesthesia
When stimulation of one of the senses causes an involuntary perception in another sense. For example, perceiving specific colours when seeing or thinking about specific numbers. (See 'Mirror-touch synaesthesia'.)

Synapse
A tiny gap between the ends of two nerves. Nerve signals pass across the synapse (see 'Neurotransmitter').

Thalamus
The great relay station of the brain. Located low and deep in the brain, all sensory information (apart from smell) travels into the thalamus before being sent to the relevant locations across the brain.

Transcranial direct current stimulation
A process where low electric currents are delivered (via electrodes attached to the head) to specific areas of the brain to activate them.

TENS (transcutaneous electrical nerve stimulation)
A process where a mild electric current is passed into the skin through electrodes connected to a battery-powered device, known as a TENS machine. Some people swear by it, but the evidence for its effectiveness is mixed.

Utilitarianism

An ethical theory that argues, generally speaking, that we must act to maximize the greatest pleasure and happiness for the greatest number of people.

Ventral pallidum

A small structure within the brain's basal ganglia (see 'Basal ganglia'), the ventral pallidum is a key part of the brain's reward circuits, playing roles in motivation and addiction.

ZFHX2

The oddly named 'zinc finger homeobox 2' (ZFHX2) is a little-known gene that plays a role in regulating the reading of genes. In 2018, members of the Marsili family from Tuscany, famed locally for their high pain thresholds, were identified to have a mutation in this gene that revealed new routes for possible future pain relievers.

References

Author's Note

[1] Edelstein, L., 'The Hippocratic Oath: Text, Translation and Interpretation', *Ancient Medicine: Selected Papers of Ludwig Edelstein*, eds. Temkin, R. and Lilian, C., Johns Hopkins University Press, 1967, pp.1484–5

Prologue

[1] Manchikanti, L., Singh, V., Datta, S., Cohen, S. P. and Hirsch, J. A., 'Comprehensive review of epidemiology, scope, and impact of spinal pain', *Pain Physician*, 12(4), 2009, pp.E35–70

[2] Jarvik, J. G. and Deyo, R. A., 'Diagnostic evaluation of low back pain with emphasis on imaging', *Annals of Internal Medicine*, 137(7), 2002, pp.586–97

[3] Vos, T., Abajobir, A. A., Abate, K. H. *et al.*, 'Global, regional, and national incidence, prevalence, and years lived with disability for 328 diseases and injuries for 195 countries, 1990–2016: a systematic analysis for the Global Burden of Disease Study 2016', *The Lancet*, 390(10100), 2017, pp.1211–59

1 The Ministry of Defence

[1] Fisher, J. P., Hassan, D. T. and O'Connor, N., 'Minerva', *BMJ*, 310(70), 1995

[2] Bayer, T. L., Baer, P. E. and Early, C., 'Situational and psychophysiological factors in psychologically induced pain', *Pain*, 44(1), 1991, pp.45–50

[3] Shakespeare, W., *The Merchant of Venice: Texts and Contexts*, ed. Kaplan, M. L., Palgrave Macmillan, 2002, pp.25–120

[4] Descartes, R., *Treatise of Man*, Harvard University Press, 1972

[5] Sherrington, C., 'The integrative action of the nervous system', *Journal of Nervous and Mental Disease*, 34(12), 1907, p.801

[6] Tewksbury, J. J. and Nabhan, G. P., 'Directed deterrence by capsaicin in chillies', *Nature*, 412(6845), 2001, pp.403–4

[7] Wall, P. D. and McMahon, S. B., 'The relationship of perceived pain to afferent nerve impulses', *Trends in Neurosciences*, 9(6), 1986, pp.254–5

[8] Melzack, R. and Wall, P. D., 'Pain mechanisms: a new theory', *Science*, 150(3699), 1965, pp.971–9

[9] Morton, D. L., Sandhu, J. S. and Jones, A. K., 'Brain imaging of pain: state of the art', *Journal of Pain Research*, 9, 2016, p.613

[10] Raja, S. N., Carr, D. B., Cohen, M. *et al.*, 'The revised International Association for the Study of Pain definition of pain: concepts, challenges, and compromises', *Pain*, 161(9), pp.1976–82

[11] Ramachandran, V. S. and Blakeslee, S., *Phantoms in the Brain: Probing the Mysteries of the Human Mind*, William Morrow, 1998, p.224

[12] Adelson, E. H., 'Checker shadow illusion', 1995

[13] MacKay, D. M., 'The epistemological problem for automata', *Automata Studies*, 1956, pp.235–52

[14] Beecher, H. K., 'Relationship of significance of wound to pain experienced', *Journal of the American Medical Association*, 161(17), 1956, pp.1609–13

2 The Painless Five

[1] Knight, T., 'Bacon: The Slice of Life', *The Kitchen As Laboratory: Reflections on the Science of Food and Cooking*, Columbia University Press, 2012, pp.73–82

[2] Dearborn, G. V. N., 'A case of congenital general pure analgesia', *Journal of Nervous and Mental Disease*, 75, 1932, pp.612–15

[3] Cox J. J., Reimann, F., Nicholas, A. K. *et al.*, 'An SCN9A channelopathy causes congenital inability to experience pain', *Nature*, 444(7121), 2006, pp.894–8

[4] McDermott, L. A., Weir, G. A., Themistocleous, A. C. *et al.*, 'Defining the functional role of $Na_v1.7$ in human nociception', *Neuron*, 101(5), 2019, pp.905–19

[5] Minett, M. S., Pereira, V., Sikandar, S. *et al.*, 'Endogenous opioids contribute to insensitivity to pain in humans and mice lacking sodium channel $Na_v1.7$', *Nature Communications*, 6(8967), 2015

[6] Fertleman, C. R., Baker, M .D., Parker, K. A. *et al.*, 'SCN9A mutations in paroxysmal extreme pain disorder: allelic variants underlie distinct channel defects and phenotypes', *Neuron*, 52(5), 2006, pp.767–74

[7] Moyer, B. D., Murray, J. K., Ligutti, J. *et al.*, 'Pharmacological characterization of potent and selective $Na_v1.7$ inhibitors engineered from Chilobrachys jingzhao tarantula venom peptide JzTx-V', *PLOS ONE*, 13(5), 2018, p.e0196791

[8] Woods, C. G., Babiker, M. O. E., Horrocks, I., Tolmie, J. and Kurth, I., 'The phenotype of congenital insensitivity to pain due to the $Na_v1.9$ variant p.L811P', *European Journal of Human Genetics*, 23, 2015, pp.561–3

[9] Habib, A. M., Matsuyama, A., Okorokov, A. L. *et al.*, 'A novel human pain insensitivity disorder caused by a point mutation in ZFHX2', *Brain*, 141(2), 2018, pp.365–76

[10] Sasso, O., Pontis, S., Armirotti, A. *et al.*, 'Endogenous *N*-acyl taurines regulate skin wound healing', *Proceedings of the National Academy of Sciences*, 113(30), 2016, pp.E4397–406

[11] Bluett, R. J., Báldi, R., Haymer, A. *et al.*, 'Endocannabinoid signalling modulates susceptibility to traumatic stress exposure', *Nature Communications*, 8(14782), 2017, pp.1–18

[12] Van Esbroeck, A. C., Janssen, A. P., Cognetta, A. B. *et al.*, 'Activity-based protein profiling reveals off-target proteins of the FAAH inhibitor BIA 10-2474', *Science*, 356(6342), 2017, pp.1084–7

[13] Lee, M. C., Nahorski, M. S., Hockley, J. R. *et al.*, 'Human labor pain is influenced by the voltage-gated potassium channel $K_v6.4$ subunit', *Cell Reports*, 32(3), 2020, p.107941

[14] Andresen, T., Lunden, D., Drewes, A. M. and Arendt-Nielsen, L., 'Pain sensitivity and experimentally induced sensitisation in red haired females', *Scandinavian Journal of Pain*, 2(1), 2011, pp.3–6

[15] Wienemann, T., Chantelau, E. A. and Koller, A., 'Effect of painless diabetic neuropathy on pressure pain hypersensitivity (hyperalgesia) after acute foot trauma', *Diabetic Foot & Ankle*, 5(1), 2014, p.24926

[16] Ndosi, M., Wright-Hughes, A., Brown, S. *et al.*, 'Prognosis of the infected diabetic foot ulcer: a 12-month prospective observational study', *Diabetic Medicine*, 35(1), 2018, pp.78–88

[17] Roglic, G., 'WHO Global report on diabetes: A summary', *International Journal of Noncommunicable Diseases*, 1(1), 2016, p.3

[18] Pop-Busui, R., Lu, J., Lopes, N. and Jones, T. L., 'Prevalence of diabetic peripheral neuropathy and relation to glycemic control therapies at baseline in the BARI 2D cohort', *Journal of the Peripheral Nervous System*, 14(1), 2009, pp.1–13

[19] Narres M., Kvitkina, T., Claessen H. *et al.*, 'Incidence of lower extremity amputations in the diabetic compared with the non-diabetic population: A systematic review', *PLOS ONE*, 12(8), 2017, p.e0182081

[20] Kerr, M., Barron, E., Chadwick, P. *et al.*, 'The cost of diabetic foot ulcers and amputations to the National Health Service in England', *Diabetic Medicine*, 36(8), 2019, pp.995–1002

[21] Schilder, P. and Stengel, E., 'Asymbolia for pain', *Archives of Neurology & Psychiatry*, 25(3), 1931, pp.598–600

[22] Berthier, M., Starkstein, S. and Leiguarda, R., 'Asymbolia for pain: a sensory–limbic disconnection syndrome', *Annals of Neurology: Official Journal of the American Neurological Association and the Child Neurology Society*, 24(1), 1988, pp.41–9

[23] Hagiwara, K., Garcia-Larrea, L., Tremblay, L. *et al.*, 'Pain behavior without pain sensation: an epileptic syndrome of "symbolism for pain"?', *Pain*, 161(3), 2020, pp.502–8

[24] Ploner, M., Freund, H. J. and Schnitzler, A., 'Pain affect without pain sensation in a patient with a postcentral lesion', *Pain*, 81(1–2), 1999, pp.211–14

3 Do I Have Your Attention?

[1] Hoffman, H. G., Chambers, G. T., Meyer III, W. J. *et al.*, 'Virtual reality as an adjunctive non-pharmacologic analgesic for acute burn pain during medical procedures', *Annals of Behavioral Medicine*, 41(2), pp.183–91

[2] Maani, C. V., Hoffman, H. G., Fowler, M. *et al.*, 'Combining ketamine and virtual reality pain control during severe burn wound care: one military and one civilian patient', *Pain Medicine*, 12(4), 2011, pp.673–8

[3] Mallari, B., Spaeth, E. K., Goh, H. and Boyd, B. S., 'Virtual reality as an analgesic for acute and chronic pain in adults: a systematic review and meta-analysis', *Journal of Pain Research*, 12, 2019, pp.2053–85

[4] 'Paget, Henry William, First Marquess of Anglesey (1768–1854), Army Officer and Politician', *Oxford Dictionary of National Biography*, Oxford University Press, 2004 (online edition)

[5] Titus Lucretius Carus, *Lucretius: The Nature of Things*, trans. Stallings, A. E., Penguin Classics, 2007

[6] Hall, K. R. L. and Stride, E., 'The varying response to pain in psychiatric disorders: a study in abnormal psychology', *British Journal of Medical Psychology*, 27(1–2), 1954, pp.48–60

[7] Sprenger, C., Eippert, F., Finsterbusch, J., Bingel, U., Rose, M. and Büchel, C., 'Attention modulates spinal cord responses to pain', *Current Biology*, 22(11), 2012, pp.1019–22

[8] Herr, H. W., 'Franklin, Lavoisier, and Mesmer: origin of the controlled clinical trial', *Urologic Oncology: Seminars and Original Investigations*, 23(5), 2005, pp.346–51

[9] Flik, C. E., Laan, W., Zuithoff, N. P. *et al.*, 'Efficacy of individual and group hypnotherapy in irritable bowel syndrome (IMAGINE): a multicentre randomised controlled trial', *The Lancet Gastroenterology & Hepatology*, 4(1), 2019, pp.20–31

[10] Miller, V., Carruthers, H. R., Morris, J., Hasan, S. S., Archbold, S. and Whorwell, P. J., 'Hypnotherapy for irritable bowel syndrome: an audit of one thousand adult patients', *Alimentary Pharmacology & Therapeutics*, 41(9), 2015, pp.844–55

[11] McGlashan, T. H., Evans, F. J. and Orne, M. T., 'The nature of hypnotic analgesia and placebo response to experimental pain', *Psychosomatic Medicine*, 31(3), 1969, pp.227–46

[12] Hilgard, E. R., 'A neodissociation interpretation of pain reduction in hypnosis', *Psychological Review*, 80(5), 1973, p.396–411

[13] Kosslyn, S. M., Thompson, W. L., Costantini-Ferrando, M. F., Alpert, N. M. and Spiegel, D., 'Hypnotic visual illusion alters color processing in the brain', *American Journal of Psychiatry*, 157(8), 2000, pp.1279–84

[14] Jiang, H., White, M. P., Greicius, M. D., Waelde, L. C. and Spiegel, D., 'Brain activity and functional connectivity associated with hypnosis', *Cerebral Cortex*, 27(8), 2017, pp.4083–93

[15] Schulz-Stübner, S., Krings, T., Meister, I. G., Rex, S., Thron, A. and Rossaint, R., 'Clinical hypnosis modulates functional magnetic resonance imaging signal intensities and pain perception in a thermal stimulation paradigm', *Regional Anesthesia & Pain Medicine*, 29(6), 2004, pp.549–56

[16] Rainville, P., Carrier, B., Hofbauer, R. K., Bushnell, M. C. and Duncan, G. H., 'Dissociation of sensory and affective dimensions of pain using hypnotic modulation', *Pain*, 82(2), 1999, pp.159–71

[17] Flik, C. E., Laan, W., Zuithoff, N. P. *et al.*, 'Efficacy of individual and group hypnotherapy in irritable bowel syndrome (IMAGINE): a multicentre randomised controlled trial', *The Lancet Gastroenterology & Hepatology*, 4(1), 2019, pp.20–31

[18] Butler, L. D., Koopman, C., Neri, E. *et al.*, 'Effects of supportive-expressive group therapy on pain in women with metastatic breast cancer', *Health Psychology*, 28(5), 2009, pp.579–87

[19] Accardi, M. C. and Milling, L. S., 'The effectiveness of hypnosis for reducing procedure-related pain in children and adolescents: a comprehensive methodological review', *Journal of Behavioral Medicine*, 32(4), 2009, pp.328–39

[20] Berlière, M., Roelants, F., Watremez *et al.*, 'The advantages of hypnosis intervention on breast cancer surgery and adjuvant therapy', *The Breast*, 37, 2018, pp.114–118

[21] Lang, E. V., Berbaum, K. S., Faintuch, S. *et al.*, 'Adjunctive self-hypnotic relaxation for outpatient medical procedures: a prospective randomized trial with women undergoing large core breast biopsy', *Pain*, 126(1–3), 2006, pp.155–64

[22] Landolt, A. S. and Milling, L. S., 'The efficacy of hypnosis as an intervention for labor and delivery pain: a comprehensive methodological review', *Clinical Psychology Review*, 31(6), 2011, pp.1022–31

[23] Vlieger, A. M., Rutten, J. M., Govers, A. M., Frankenhuis, C. and Benninga, M. A., 'Long-term follow-up of gut-directed hypnotherapy vs. standard care in children with functional abdominal pain or irritable bowel syndrome', *American Journal of Gastroenterology*, 107(4), 2012, pp.627–31

[24] Jensen, M. P., Mendoza, M. E., Ehde, D. M. *et al.*, 'Effects of hypnosis, cognitive therapy, hypnotic cognitive therapy, and pain education in adults with chronic pain: a randomized clinical trial', *Pain*, 161(10), 2020, pp.2284–98

[25] Larbig, W., Elbert, T., Lutzenberger, W., Rockstroh, B., Schnerr, G. and Birbaumer, N., 'EEG and slow brain potentials during anticipation and control of painful stimulation', *Electroencephalography and Clinical Neurophysiology*, 53(3), 1982, pp.298–309

[26] Jensen, M. P., Adachi, T. and Hakimian, S., 'Brain oscillations, hypnosis, and hypnotizability', *American Journal of Clinical Hypnosis*, 57(3), 2015, pp.230–53

[27] Guilbert, A. S., Chauvin, C. and De Melo, C., 'Effect of virtual reality hypnosis on postoperative pain and morphine consumption after surgery for scoliosis: a retrospective evaluation in children', abstract A2375 from the Anesthesiology Annual Meeting, 2018

4 The Expectation Effect

[1] 'Headaches, chilli pepper patches and the placebo effect', *Airing Pain, 53*, painconcern.org.uk, 30 January 2014

[2] Chaucer, G., *The Canterbury Tales*, eds. Boenig, R. and Taylor, A., Broadview Press, 2012

[3] Handfield-Jones, R. P. C., 'A bottle of medicine from the doctor', *The Lancet*, 262(6790), 1953, pp.823–25

[4] Hróbjartsson, A. and Gøtzsche, P. C., 'Is the placebo powerless? An analysis of clinical trials comparing placebo with no treatment', *New England Journal of Medicine*, 344(21), 2001, pp.1594–1602

[5] Moseley, J. B., O'Malley, K., Petersen, N. J. *et al.*, 'A controlled trial of arthroscopic surgery for osteoarthritis of the knee', *New England Journal of Medicine*, 347(2), pp.81–8

[6] Thorlund, J. B., Juhl, C. B., Roos, E. M. and Lohmander, L. S., 'Arthroscopic surgery for degenerative knee: systematic review and meta-analysis of benefits and harms', *BMJ*, 350, 2015, p.h2747

[7] Wartolowska, K., Judge, A., Hopewell, S. *et al.*, 'Use of placebo controls in the evaluation of surgery: systematic review', *BMJ*, 348, 2014

[8] Wager, T. D., Rilling, J. K., Smith, E. E. *et al.*, 'Placebo-induced changes in FMRI in the anticipation and experience of pain', *Science*, 303(5661), 2004, pp.1162–7

[9] Wager, T. D., Scott, D. J. and Zubieta, J. K., 'Placebo effects on human μ-opioid activity during pain', *Proceedings of the National Academy of Sciences*, 104(26), 2007, pp.11056–61

[10] Levine, J., Gordon, N. and Fields, H., 'The mechanism of placebo analgesia', *The Lancet*, 312(8091), 1978, pp.654–7

[11] Eippert, F., Bingel, U., Schoell, E. D. *et al.*, 'Activation of the opioidergic descending pain control system underlies placebo analgesia', *Neuron*, 63(4), pp.533–43

[12] Benedetti, F., Amanzio, M., Rosato, R. and Blanchard, C., 'Nonopioid placebo analgesia is mediated by CB1 cannabinoid receptors', *Nature Medicine*, 17(10), 2011, pp.1228–30

[13] Scott, D. J., Stohler, C. S., Egnatuk, C. M., Wang, H., Koeppe, R. A. and Zubieta, J. K., 'Individual differences in reward responding explain placebo-induced expectations and effects', *Neuron*, 55(2), 2007, pp.325–36

[14] Eippert, F., Finsterbusch, J., Bingel, U. and Büchel, C., 'Direct evidence for spinal cord involvement in placebo analgesia', *Science*, 326(5951), 2009, p.404

[15] Bannuru, R. R., McAlindon, T. E., Sullivan, M. C., Wong, J. B., Kent, D. M. and Schmid, C.H., 'Effectiveness and implications of alternative placebo treatments: a systematic review and network meta-analysis of osteoarthritis trials', *Annals of Internal Medicine*, 163(5), 2015, pp.365–72

[16] Espay, A. J., Norris, M. M., Eliassen, J. C. *et al.*, 'Placebo effect of medication cost in Parkinson disease: a randomized double-blind study', *Neurology*, 84(8), 2015, pp.794–802

[17] Haake, M., Müller, H. H., Schade-Brittinger, C. *et al.*, 'German acupuncture trials (GERAC) for chronic low back pain: randomized, multicenter, blinded, parallel-group trial with 3 groups', *Archives of Internal Medicine*, 167(17), 2007, pp.1892–8

[18] Tuttle, A. H., Tohyama, S., Ramsay, T. *et al.*, 'Increasing placebo responses over time in US clinical trials of neuropathic pain', *Pain*, 156(12), 2015, pp.2616–26

[19] Amanzio, M., Pollo, A., Maggi, G. and Benedetti, F., 'Response variability to analgesics: a role for non-specific activation of endogenous opioids', *Pain*, 90(3), 2001, pp.205–15

[20] Gracely, R. H., Dubner, R., Deeter, W. R. and Wolskee, P. J., 'Clinicians' expectations influence placebo analgesia', *The Lancet*, 1(8419), 1985

[21] Morton, D. L., Watson, A., El-Deredy, W. and Jones, A. K., 'Reproducibility of placebo analgesia: effect of dispositional optimism', *Pain*, 146(1–2), 2009, pp.194–8

[22] Barsky, A. J., Saintfort, R., Rogers, M. P. and Borus, J. F., 'Nonspecific medication side effects and the nocebo phenomenon', *JAMA*, 287(5), 2002, pp.622–7

[23] Wood, F. A., Howard, J. P., Finegold, J. A. *et al.*, 'N-of-1 trial of a statin, placebo, or no treatment to assess side effects', *New England Journal of Medicine*, 383, 2020, pp.2182–4

[24] Bartholomew, R. E. and Wessely, S., 'Protean nature of mass sociogenic illness: from possessed nuns to chemical and biological terrorism fears', *British Journal of Psychiatry*, 180(4), 2002, pp.300–6

[25] Benedetti, F., Lanotte, M., Lopiano, L. and Colloca, L., 'When words are painful: unraveling the mechanisms of the nocebo effect', *Neuroscience*, 147(2), 2007, pp.260–71

[26] Ritter, A., Franz, M., Puta, C., Dietrich, C., Miltner, W. H. and Weiss, T., 'Enhanced brain responses to pain-related words in chronic back pain patients and their modulation by current pain', *Healthcare*, 4(3), 2016, p.54

[27] Hansen, E. and Zech, N., 'Nocebo effects and negative suggestions in daily clinical practice – forms, impact and approaches to avoid them', *Frontiers in Pharmacology*, 10, 2019, p.77

[28] Varelmann, D., Pancaro, C., Cappiello, E. C. and Camann, W. R., 'Nocebo-induced hyperalgesia during local anesthetic injection', *Anesthesia & Analgesia*, 110(3), 2010, pp.868–70

[29] Bingel, U., Wanigasekera, V., Wiech, K. *et al.*, 'The effect of treatment expectation on drug efficacy: imaging the analgesic benefit of the opioid remifentanil', *Science Translational Medicine*, 3(70), 2011, p.70ra14

[30] Amanzio, M., Pollo, A., Maggi, G. and Benedetti, F., 'Response variability to analgesics: a role for non-specific activation of endogenous opioids', *Pain*, 90(3), 2001, pp.205–15

[31] Walach, H. and Jonas, W. B., 'Placebo research: the evidence base for harnessing self-healing capacities', *Journal of Alternative & Complementary Medicine*, 10 (Supplement 1), 2004, p.S-103

[32] Interview with Dan Moerman in Marchant, J., *Cure: A Journey into the Science of Mind Over Body*, Broadway Books, 2016

[33] Conboy, L. A., Macklin, E., Kelley, J., Kokkotou, E., Lembo, A. and Kaptchuk, T., 'Which patients improve: characteristics increasing sensitivity to a supportive patient–practitioner relationship', *Social Science & Medicine*, 70(3), 2010, pp.479–84

[34] Ernst, E., 'A systematic review of systematic reviews of homeopathy', *British Journal of Clinical Pharmacology*, 54(6), 2002, pp.577–82

[35] Specter, M., 'The power of nothing', *New Yorker*, 5 December 2011

[36] Kaptchuk, T. J., Friedlander, E., Kelley, J. M. *et al.*, 'Placebos without deception: a randomized controlled trial in irritable bowel syndrome', *PLOS ONE*, 5(12), 2010, p.e15591

[37] Carvalho, C., Caetano, J. M., Cunha, L., Rebouta, P., Kaptchuk, T. J. and Kirsch, I., 'Open-label placebo treatment in chronic low back pain: a randomized controlled trial', *Pain*, 157(12), 2016, p.2766–72

[38] Kam-Hansen, S., Jakubowski, M., Kelley, J. M. *et al.*, 'Altered placebo and drug labeling changes the outcome of episodic migraine attacks', *Science Translational Medicine*, 6(218), 2014, p.218ra5

[39] Wang, R. S., Hall, K. T., Giulianini, F., Passow, D., Kaptchuk, T. J. and Loscalzo, J., 'Network analysis of the genomic basis of the placebo effect', *JCI Insight*, 2(11), 2017, p.e93911

[40] Colloca, L. and Benedetti, F., 'How prior experience shapes placebo analgesia', *Pain*, 124(1–2), 2006, pp.126–33

[41] Schafer, S. M., Colloca, L. and Wager, T. D., 'Conditioned placebo analgesia persists when subjects know they are receiving a placebo', *Journal of Pain*, 16(5), 2015, pp.412–20

[42] Tu, Y., Park, J., Ahlfors, S. P. *et al.*, 'A neural mechanism of direct and observational conditioning for placebo and nocebo responses', *NeuroImage*, 184, 2019, pp.954–63

[43] Colloca, L., Enck, P. and DeGrazia, D., 'Relieving pain using dose-extending placebos: a scoping review', *Pain*, 157(8), 2016, pp.1590–98

[44] Thompson, P., 'Margaret Thatcher: A new illusion', *Perception*, 9(4), 1980, pp.483–4

[45] Summerfield, C., Egner, T., Greene, M., Koechlin, E., Mangels, J. and Hirsch, J., 'Predictive codes for forthcoming perception in the frontal cortex', *Science*, 314(5803), 2006, pp.1311–14

[46] George, K. and Das, J. M., 'Neuroanatomy, thalamocortical radiations', StatPearls Publishing, 2019

[47] Wallisch, P., 'Illumination assumptions account for individual differences in the perceptual interpretation of a profoundly ambiguous stimulus in the color domain: "The dress"', *Journal of Vision*, 17(4), 2017

[48] Casey, K., 'Theory of predictive brain as important as evolution – Prof. Lars Muckli', *Horizon*, 29 May 2018

[49] Ongaro, G. and Kaptchuk, T. J., 'Symptom perception, placebo effects, and the Bayesian brain', *Pain*, 160(1), 2019, pp.1–4

[50] Kaptchuk, T. J., 'Open-label placebo: reflections on a research agenda', *Perspectives in Biology and Medicine*, 61(3), 2018, pp.311–34

5 The Meaning of Pain

[1] International Committee of the Red Cross (ICRC), Geneva Convention Relative to the Protection of Civilian Persons in Time of War (Fourth Geneva Convention), 12 August 1949, 75 UNTS 287

[2] Tsur, N., Defrin, R. and Ginzburg, K., 'Posttraumatic stress disorder, orientation to pain, and pain perception in ex-prisoners of war who underwent torture', *Psychosomatic Medicine*, 79(6), 2017, pp.655–63

[3] Raja, S. N., Carr, D. B., Cohen, M. *et al.*, 'The revised International Association for the Study of Pain definition of pain: concepts, challenges, and compromises', *Pain*, 161(9), 2020, pp.1976–82

[4] Shackman, A. J. and Wager, T. D., 'The emotional brain: fundamental questions and strategies for future research', *Neuroscience Letters*, 693, 2019, pp.68–74

[5] Eisenberger, N. I., Lieberman, M. D. and Williams, K. D., 'Does rejection hurt? An fMRI study of social exclusion', *Science*, 302(5643), 2003, pp.290–2

[6] DeWall, C. N., MacDonald, G., Webster, G. D. *et al.*, 'Acetaminophen reduces social pain: behavioral and neural evidence', *Psychological Science*, 21(7), 2010, pp.931–7

[7] Ratner, K. G., Kaczmarek, A. R. and Hong, Y., 'Can over-the-counter pain medications influence our thoughts and emotions?', *Policy Insights from the Behavioral and Brain Sciences*, 5(1), 2018, pp.82–9

[8] Farrell, S. M., Green, A. and Aziz, T., 'The current state of deep brain stimulation for chronic pain and its context in other forms of neuromodulation', *Brain Sciences*, 8(8), 2018, p.158

[9] Lempka, S. F., Malone Jr, D. A., Hu, B. *et al.*, 'Randomized clinical trial of deep brain stimulation for poststroke pain', *Annals of Neurology*, 81(5), 2017, pp.653–63

[10] Ploghaus, A., Narain, C., Beckmann, C.F. *et al.*, 'Exacerbation of pain by anxiety is associated with activity in a hippocampal network', *Journal of Neuroscience*, 21(24), 2001, pp.9896–9903

[11] Zhou, F., Shefer, A., Wenger, J. *et al.*, 'Economic evaluation of the routine childhood immunization program in the United States, 2009', *Pediatrics*, 133(4), 2014, pp.577–85

[12] McMurtry, C. M., Riddell, R. P., Taddio, A. *et al.*, 'Far from "just a poke": common painful needle procedures and the development of needle fear', *Clinical Journal of Pain*, 31 (Supplement 10), 2015, pp.S3–11

[13] Taddio, A., McMurtry, C. M., Shah, V. *et al.*, 'Reducing pain during vaccine injections: clinical practice guideline', *CMAJ*, 187(13), 2015, pp.975–82

[14] Wang, Y., Wang, J. Y. and Luo, F., 'Why self-induced pain feels less painful than externally generated pain: distinct brain activation patterns in self- and externally generated pain', *PLOS ONE*, 6(8), 2011, p.e23536

[15] Mowrer, O. H. and Viek, P., 'An experimental analogue of fear from a sense of helplessness', *Journal of Abnormal and Social Psychology*, 43(2), 1948, pp.193–200

[16] Bowers, K. S., 'Pain, anxiety, and perceived control', *Journal of Consulting and Clinical Psychology*, 32(5) (Part 1), 1968, pp.596–602

[17] Segal, Z. V., Kennedy, S., Gemar, M., Hood, K., Pedersen, R. and Buis, T., 'Cognitive reactivity to sad mood provocation and the prediction of depressive relapse', *Archives of General Psychiatry*, 63(7), 2006, pp.749–55

[18] Berna, C., Leknes, S., Holmes, E. A., Edwards, R. R., Goodwin, G. M. and Tracey, I., 'Induction of depressed mood disrupts emotion regulation neurocircuitry and enhances pain unpleasantness', *Biological Psychiatry*, 67(11), 2010, pp.1083–90

[19] Andersson, G. B., 'Epidemiological features of chronic low-back pain', *The Lancet*, 354(9178), 1999, pp.581–5

[20] Vlaeyen, J. W. and Linton, S. J., 'Fear-avoidance and its consequences in chronic musculoskeletal pain: a state of the art', *Pain*, 85(3), 2000, pp. 317–32

[21] Hashmi, J. A., Baliki, M. N., Huang, L. *et al.*, 'Shape shifting pain: chronification of back pain shifts brain representation from nociceptive to emotional circuits', *Brain*, 136(Part 9), 2013, pp.2751–68

[22] Price, D. D., 'Psychological and neural mechanisms of the affective dimension of pain', *Science*, 288(5472), 2000, pp.1769–72

[23] Wertli, M. M., Burgstaller, J. M., Weiser, S., Steurer, J., Kofmehl, R. and Held, U., 'Influence of catastrophizing on treatment outcome in patients with nonspecific low back pain: a systematic review', *Spine*, 39(3), 2014, pp.263–73

[24] Cherkin, D. C., Sherman, K. J., Balderson, B. H. *et al.*, 'Effect of mindfulness-based stress reduction vs cognitive behavioral therapy or usual care on back pain and functional limitations in adults with chronic low back pain: a randomized clinical trial', *JAMA*, 315(12), 2016, pp.1240–9

[25] Hughes, L. S., Clark, J., Colclough, J. A., Dale, E. and McMillan, D., 'Acceptance and commitment therapy (ACT) for chronic pain', *Clinical Journal of Pain*, 33(6), 2017, pp.552–68

[26] Lutz, A., McFarlin, D. R., Perlman, D. M., Salomons, T. V. and Davidson, R. J., 'Altered anterior insula activation during anticipation and experience of painful stimuli in expert meditators', *NeuroImage*, 64, 2013, pp.538–46

[27] Lumley, M. A., Schubiner, H., Lockhart, N. A. *et al.*, 'Emotional awareness and expression therapy, cognitive-behavioral therapy, and education for fibromyalgia: a cluster-randomized controlled trial', *Pain*, 158(12), 2017, pp.2354–63

[28] Lumley, M. A. and Schubiner, H., 'Psychological therapy for centralized pain: an integrative assessment and treatment model', *Psychosomatic Medicine*, 81(2), 2019, pp.114–24

[29] C de C Williams, A., Fisher, E., Hearn L. and Eccleston, C., 'Psychological therapies for the management of chronic pain (excluding headache) in adults', *Cochrane Database of Systematic Reviews*, 8, 2020, CD007407

6 No Pain, No Gain

[1] Bentham, J., *The Principles of Morals and Legislation*, Prometheus Books, 1988, pp.57–79

[2] Leknes, S., Berna, C., Lee, M. C., Snyder, G. D., Biele, G. and Tracey, I., 'The importance of context: when relative relief renders pain pleasant', *Pain*, 154(3), 2013, pp.402–10

[3] Ameriks, K. and Clarke, D. M., *Aristotle: Nicomachean Ethics*, Cambridge University Press, 2000

[4] Price, D. D., Harkins, S. W. and Baker, C., 'Sensory-affective relationships among different types of clinical and experimental pain', *Pain*, 28(3), 1987, pp.297–307

[5] Petrovic, P., Dietrich, T., Fransson, P., Andersson, J., Carlsson, K. and Ingvar, M., 'Placebo in emotional processing—induced expectations of anxiety relief activate a generalized modulatory network', *Neuron*, 46(6), 2005, pp.957–69

[6] Harper, P., 'No pain, no gain: pain behaviour in the armed forces', *British Journal of Nursing*, 15(10), 2006, pp.548–51

[7] Fields, H. L., 'A motivation-decision model of pain: the role of opioids', *Proceedings of the 11th World Congress on Pain*, IASP Press, 2006

[8] Barbano, M. F. and Cador, M., 'Differential regulation of the consummatory, motivational and anticipatory aspects of feeding behavior by dopaminergic and opioidergic drugs', *Neuropsychopharmacology*, 31(7), 2006, pp.1371–81

[9] Forsberg, G., Wiesenfeld-Hallin, Z., Eneroth, P. and Södersten, P., 'Sexual behavior induces naloxone-reversible hypoalgesia in male rats', *Neuroscience Letters*, 81(1–2), 1987, pp.151–4

[10] Sharot, T., Shiner, T., Brown, A. C., Fan, J. and Dolan, R. J., 'Dopamine enhances expectation of pleasure in humans', *Current Biology*, 19(24), 2009, pp.2077–80

[11] Budygin, E. A., Park, J., Bass, C. E., Grinevich, V. P., Bonin, K. D. and Wightman, R. M., 'Aversive stimulus differentially triggers subsecond dopamine release in reward regions', *Neuroscience*, 201, 2012, pp.331–7

[12] Leknes, S., Lee, M., Berna, C., Andersson, J. and Tracey, I., 'Relief as a reward: hedonic and neural responses to safety from pain', *PLOS ONE*, 6(4), 2011, p.e17870

[13] Zubieta, J. K., Heitzeg, M. M., Smith, Y. R. *et al.*, 'COMT val158met genotype affects µ-opioid neurotransmitter responses to a pain stressor', *Science*, 299(5610), 2003, pp.1240–43

[14] Durso, G. R., Luttrell, A. and Way, B. M., 'Over-the-counter relief from pains and pleasures alike: acetaminophen blunts evaluation sensitivity to both negative and positive stimuli', *Psychological Science*, 26(6), 2015, pp.750–8

[15] Forsberg, G., Wiesenfeld-Hallin, Z., Eneroth, P. and Södersten, P., 'Sexual behavior induces naloxone-reversible hypoalgesia in male rats', *Neuroscience Letters*, 81(1–2), 1987, pp.151–4

[16] Roy, M., Peretz, I. and Rainville, P., 'Emotional valence contributes to music-induced analgesia', *Pain*, 134(1–2), 2008, pp.140–7

[17] Gandhi, W. and Schweinhardt, P., 'How accurate appraisal of behavioral costs and benefits guides adaptive pain coping', *Frontiers in Psychiatry*, 8, 2017, p.103

[18] Baliki, M. N., Petre, B., Torbey, S. *et al.*, 'Corticostriatal functional connectivity predicts transition to chronic back pain', *Nature Neuroscience*, 15(8), 2012, pp.1117–19

[19] Kaneko, H., Zhang, S., Sekiguchi, M. *et al.*, 'Dysfunction of nucleus accumbens is associated with psychiatric problems in patients with chronic low back pain: a functional magnetic resonance imaging study', *Spine*, 42(11), 2017, pp.844–53

[20] Taylor, A. M., Becker, S., Schweinhardt, P. and Cahill, C., 'Mesolimbic dopamine signaling in acute and chronic pain: implications for motivation, analgesia, and addiction', *Pain*, 157(6), 2016, p.1194

[21] Loggia, M. L., Berna, C., Kim, J. *et al.*, 'Disrupted brain circuitry for pain-related reward/punishment in fibromyalgia', *Arthritis & Rheumatology*, 66(1), 2014, pp.203–12

[22] Rozin, P., Guillot, L., Fincher, K., Rozin, A. and Tsukayama, E., 'Glad to be sad, and other examples of benign masochism', *Judgment and Decision Making*, 8(4), 2013, pp.439–47

[23] McGraw, A. P., Warren, C., Williams, L. E. and Leonard, B., 'Too close for comfort, or too far to care? Finding humor in distant tragedies and close mishaps', *Psychological Science*, 23(10), 2012, pp.1215–23

[24] Franklin, J. C., Lee, K. M., Hanna, E. K. and Prinstein, M. J., 'Feeling worse to feel better: pain-offset relief simultaneously stimulates positive affect and reduces negative affect', *Psychological Science*, 24(4), 2013, pp.521–9

[25] Glenn, J. J., Michel, B. D., Franklin, J. C., Hooley, J. M. and Nock, M. K., 'Pain analgesia among adolescent self-injurers', *Psychiatry Research*, 220(3), 2014, pp.921–6

[26] Kirtley, O. J., O'Carroll, R. E. and O'Connor, R. C., 'Pain and self-harm: a systematic review', *Journal of Affective Disorders*, 203, 2016, pp.347–63

[27] Fox, K. R., O'Sullivan, I. M., Wang, S. B. and Hooley, J. M., 'Self-criticism impacts emotional responses to pain', *Behavior Therapy*, 50(2), 2019, pp.410–20

[28] Niedtfeld, I., Schulze, L., Kirsch, P., Herpertz, S. C., Bohus, M. and Schmahl, C., 'Affect regulation and pain in borderline personality disorder: a possible link to the understanding of self-injury', *Biological Psychiatry*, 68(4), 2010, pp.383–91

[29] Hooley, J. M. and Franklin, J. C., 'Why do people hurt themselves? A new conceptual model of nonsuicidal self-injury', *Clinical Psychological Science*, 6(3), 2018, pp.428–51

[30] Hooley, J. M., Dahlgren, M. K., Best, S. G., Gonenc, A. and Gruber, S. A., 'Decreased amygdalar activation to NSSI-stimuli in people who engage in NSSI: a neuroimaging pilot study', *Frontiers in Psychiatry*, 11, 2020, p.238

[31] Hooley, J. M. and St. Germain, S. A., 'Nonsuicidal self-injury, pain, and self-criticism: does changing self-worth change pain endurance in people who engage in self-injury?', *Clinical Psychological Science*, 2(3), 2014, pp.297–305

7 I Feel Your Pain

[1] Salinas, J., *Mirror Touch: A Memoir of Synesthesia and the Secret Life of the Brain*, HarperCollins, 2017

[2] Miller, L. and Spiegel, A., 'Entanglement', *Invisibilia* podcast, 20 January 2015

[3] Ward, J., Schnakenberg, P. and Banissy, M. J., 'The relationship between mirror-touch synaesthesia and empathy: new evidence and a new screening tool', *Cognitive Neuropsychology*, 35(5–6), 2018, pp.314–32

[4] Banissy, M. J., Kadosh, R. C., Maus, G. W., Walsh, V. and Ward, J., 'Prevalence, characteristics and a neurocognitive model of mirror-touch synaesthesia', *Experimental Brain Research*, 198(2–3), 2009, pp.261–72

[5] Blakemore, S. J., Bristow, D., Bird, G., Frith, C. and Ward, J., 'Somatosensory activations during the observation of touch and a case of vision–touch synaesthesia', *Brain*, 128(7), 2005, pp.1571–83

[6] Goller, A. I., Richards, K., Novak, S. and Ward, J., 'Mirror-touch synaesthesia in the phantom limbs of amputees', *Cortex*, 49(1), 2013, pp.243–51

[7] Lamm, C., Decety, J. and Singer, T., 'Meta-analytic evidence for common and distinct neural networks associated with directly experienced pain and empathy for pain', *NeuroImage*, 54(3), 2011, pp.2492–502

[8] Bekkali, S., Youssef, G. J., Donaldson, P. H., Albein-Urios, N., Hyde, C. and Enticott, P. G., 'Is the putative mirror neuron system associated with empathy? A systematic review and meta-analysis', *Neuropsychology Review*, 2020, pp.1–44

[9] Rütgen, M., Seidel, E. M., Silani, G. *et al.*, 'Placebo analgesia and its opioidergic regulation suggest that empathy for pain is grounded in self pain', *Proceedings of the National Academy of Sciences*, 112(41), 2015, pp.E5638–46

[10] Decety, J., Michalska, K. J. and Akitsuki, Y., 'Who caused the pain? An fMRI investigation of empathy and intentionality in children', *Neuropsychologia*, 46(11), 2008, pp.2607–14

[11] Decety, J. and Michalska, K. J., 'Neurodevelopmental changes in the circuits underlying empathy and sympathy from childhood to adulthood', *Developmental Science*, 13(6), 2010, pp.886–99

[12] Marsh, A. A., Finger, E. C., Fowler, K. A. *et al.*, 'Empathic responsiveness in amygdala and anterior cingulate cortex in youths with psychopathic traits', *Journal of Child Psychology and Psychiatry*, 54(8), 2013, pp.900–10

[13] Lockwood, P. L., Apps, M. A., Roiser, J. P. and Viding, E., 'Encoding of vicarious reward prediction in anterior cingulate cortex and relationship with trait empathy', *Journal of Neuroscience*, 35(40), 2015, pp.13720–7

[14] Jeon, D., Kim, S., Chetana, M. *et al.*, 'Observational fear learning involves affective pain system and $Ca_v1.2$ Ca^{2+} channels in ACC', *Nature Neuroscience*, 13(4), 2010, pp.482–8

[15] Sapolsky, R. M., *Behave: The Biology of Humans at Our Best and Worst*, Penguin, 2017

[16] Decety, J., Echols, S. and Correll, J., 'The blame game: the effect of responsibility and social stigma on empathy for pain', *Journal of Cognitive Neuroscience*, 22(5), 2010, pp.985–97

[17] Xu, X., Zuo, X., Wang, X. and Han, S., 'Do you feel my pain? Racial group membership modulates empathic neural responses', *Journal of Neuroscience*, 29(26), 2009, pp.8525–9

[18] Shen, F., Hu, Y., Fan, M., Wang, H. and Wang, Z., 'Racial bias in neural response for pain is modulated by minimal group', *Frontiers in Human Neuroscience*, 11, 2018, p.661

[19] Cao, Y., Contreras-Huerta, L. S., McFadyen, J. and Cunnington, R., 'Racial bias in neural response to others' pain is reduced with other-race contact', *Cortex*, 70, 2015, pp.68–78

[20] Cikara, M. and Fiske, S. T., 'Their pain, our pleasure: stereotype content and schadenfreude', *Annals of the New York Academy of Sciences*, 1299, 2013, pp.52–9

[21] Takahashi, H., Kato, M., Matsuura, M., Mobbs, D., Suhara, T. and Okubo, Y., 'When your gain is my pain and your pain is my gain: neural correlates of envy and schadenfreude', *Science*, 323(5916), 2009, pp.937–9

[22] Singer, T., Seymour, B., O'Doherty, J. P., Stephan, K. E., Dolan, R. J. and Frith, C. D., 'Empathic neural responses are modulated by the perceived fairness of others', *Nature*, 439(7075), 2006, pp.466–9

[23] Decety, J., Yang, C. Y. and Cheng, Y., 'Physicians down-regulate their pain empathy response: an event-related brain potential study', *NeuroImage*, 50(4), 2010, pp.1676–82

[24] Lamm, C., Batson, C. D. and Decety, J., 'The neural substrate of human empathy: effects of perspective-taking and cognitive appraisal', *Journal of Cognitive Neuroscience*, 19(1), 2007, pp.42–58

[25] Klimecki, O. M., Leiberg, S., Lamm, C. and Singer, T., 'Functional neural plasticity and associated changes in positive affect after compassion training', *Cerebral Cortex*, 23(7), 2013, pp.1552–61

[26] Cánovas, L., Carrascosa, A.J., García, M. *et al.*, 'Impact of empathy in the patient–doctor relationship on chronic pain relief and quality of life: a prospective study in Spanish pain clinics', *Pain Medicine*, 19(7), 2018, pp.1304–14

[27] Gray, K., 'The power of good intentions: perceived benevolence soothes pain, increases pleasure, and improves taste', *Social Psychological and Personality Science*, 3(5), 2012, pp.639–45

[28] Butler, D. and Moseley, G., *Explain Pain Supercharged*, NOI Group, 2017.

8 Pulling Together

[1] Eisenberger, N. I., Lieberman, M. D. and Williams, K. D., 'Does rejection hurt? An fMRI study of social exclusion', *Science*, 302(5643), 2003, pp.290–2

[2] Eisenberger, N. I., Jarcho, J. M., Lieberman, M. D. and Naliboff, B. D., 'An experimental study of shared sensitivity to physical pain and social rejection', *Pain*, 126(1–3), pp.132–8

[3] Murphy, M. R., MacLean, P. D. and Hamilton, S. C., 'Species-typical behavior of hamsters deprived from birth of the neocortex', *Science*, 213(4506), 1981, pp.459–61

[4] MacLean, P. D. and Newman, J. D., 'Role of midline frontolimbic cortex in production of the isolation call of squirrel monkeys', *Brain Research*, 450(1–2), 1988, pp.111–23

[5] Martin, L. J., Tuttle, A. H. and Mogil, J. S., 'The interaction between pain and social behavior in humans and rodents', *Behavioral Neurobiology of Chronic Pain*, 2014, pp.233–50

[6] Holt-Lunstad, J., Smith, T. B. and Layton, J. B., 'Social relationships and mortality risk: a meta-analytic review', *PLOS Medicine*, 7(7), 2010, p.e1000316

[7] Karayannis, N. V., Baumann, I., Sturgeon, J. A., Melloh, M. and Mackey, S. C., 'The impact of social isolation on pain interference: a longitudinal study', *Annals of Behavioral Medicine*, 53(1), 2019, pp.65–74

[8] Cohen, E. E., Ejsmond-Frey, R., Knight, N. and Dunbar, R. I., 'Rowers' high: behavioural synchrony is correlated with elevated pain thresholds', *Biology Letters*, 6(1), 2010, pp.106–8

[9] Launay, J., Grube, M. and Stewart, L., 'Dysrhythmia: a specific congenital rhythm perception deficit', *Frontiers in Psychology*, 5, 2014, p.18

[10] Hopper, M. J., Curtis, S., Hodge, S. and Simm, R., 'A qualitative study exploring the effects of attending a community pain service choir on wellbeing in people who experience chronic pain', *British Journal of Pain*, 10(3), 2016, pp.124–34

[11] Dunbar, R. I., Baron, R., Frangou, A. *et al.*, 'Social laughter is correlated with an elevated pain threshold', *Proceedings of the Royal Society B: Biological Sciences*, 279(1731), 2012, pp.1161–7

[12] Provine, R. R. and Fischer, K. R., 'Laughing, smiling, and talking: relation to sleeping and social context in humans', *Ethology*, 83(4), 1989, pp.295–305

[13] Manninen, S., Tuominen, L., Dunbar, R. I. *et al.*, 'Social laughter triggers endogenous opioid release in humans', *Journal of Neuroscience*, 37(25), pp.6125–31

[14] Johnson, K. V. A. and Dunbar, R. I., 'Pain tolerance predicts human social network size', *Scientific Reports*, 6, 2016, p.25267

[15] Langford, D. J., Crager, S. E., Shehzad, Z. *et al.*, 'Social modulation of pain as evidence for empathy in mice', *Science*, 312(5782), 2006, pp.1967–70

[16] Goldstein, P., Shamay-Tsoory, S. G., Yellinek, S. and Weissman-Fogel, I., 'Empathy predicts an experimental pain reduction during touch', *Journal of Pain*, 17(10), 2016, pp.1049–57

[17] Huddy, J., 'A new hope: social prescribing in Cornwall', *British Journal of General Practice*, 69(682), 2019, p.243

[18] Singhal, A., Tien, Y. Y. and Hsia, R. Y., 'Racial-ethnic disparities in opioid prescriptions at emergency department visits for conditions commonly associated with prescription drug abuse', *PLOS ONE*, 11(8), 2016, p.e0159224

[19] Goyal, M. K., Kuppermann, N., Cleary, S. D., Teach, S. J. and Chamberlain, J. M., 'Racial disparities in pain management of children with appendicitis in emergency departments', *JAMA Pediatrics*, 169(11), 2015, pp.996–1002

[20] Druckman, J. N., Trawalter, S., Montes, I., Fredendall, A., Kanter, N. and Rubenstein, A.P., 'Racial bias in sport medical staff's perceptions of others' pain', *Journal of Social Psychology*, 158(6), 2018, pp.721–9

[21] Hoffman, K. M., Trawalter, S., Axt, J. R. and Oliver, M. N., 'Racial bias in pain assessment and treatment recommendations, and false beliefs about biological differences between blacks and whites', *Proceedings of the National Academy of Sciences*, 113(16), 2016, pp.4296–301

[22] Laurencin, C. T. and Murray, M., 'An American crisis: the lack of black men in medicine', *Journal of Racial and Ethnic Health Disparities*, 4(3), 2017, pp.317–21

[23] Fillingim, R. B., King, C. D., Ribeiro-Dasilva, M. C., Rahim-Williams, B. and Riley III, J. L., 'Sex, gender, and pain: a review of recent clinical and experimental findings', *Journal of Pain*, 10(5), 2009, pp.447–85

[24] Chen, E. H., Shofer, F. S., Dean, A. J. *et al.*, 'Gender disparity in analgesic treatment of emergency department patients with acute abdominal pain', *Academic Emergency Medicine*, 15(5), 2008, pp.414–18

[25] Cepeda, M. S. and Carr, D. B., 'Women experience more pain and require more morphine than men to achieve a similar degree of analgesia', *Anesthesia & Analgesia*, 97(5), 2003, pp.1464–8

[26] Bartley, E. J. and Fillingim, R. B., 'Sex differences in pain: a brief review of clinical and experimental findings', *British Journal of Anaesthesia*, 111(1), 2013, pp.52–8

[27] England, C., 'Erectile dysfunction studies outnumber PMS research by five to one', *The Independent*, 15 August 2016

[28] '10 things you should know about endometriosis', Royal College of Obstetricians and Gynaecologists, 2017

[29] Lawesson, S. S., Isaksson, R. M., Ericsson, M., Ängerud, K. and Thylén, I., 'Gender disparities in first medical contact and delay in ST-elevation myocardial infarction: a prospective multicentre Swedish survey study', *BMJ Open*, 8(5), 2018, p.e020211

[30] Moser, D. K., McKinley, S., Dracup, K. and Chung, M. L., 'Gender differences in reasons patients delay in seeking treatment for acute myocardial infarction symptoms', *Patient education and counseling*, 56(1), 2005, pp.45-54

[31] 'Naomi Musenga death: emergency operator blames pressure after mocking caller', BBC News, 14 May 2018

[32] Boseley, S., '"Listen to women": UK doctors issued with first guidance on endometriosis', *Guardian*, 6 September 2017

[33] McParland, J. L., Eccleston, C., Osborn, M. and Hezseltine, L., 'It's not fair: an interpretative phenomenological analysis of discourses of justice and fairness in chronic pain', *Health*, 15(5), 2011, pp.459–74

[34] McParland, J. L., Knussen, C. and Murray, J., 'The effects of a recalled injustice on the experience of experimentally induced pain and anxiety in relation to just-world beliefs', *European Journal of Pain*, 20(9), 2016, pp.1392–1401

[35] Trost, Z., Scott, W., Lange, J. M., Manganelli, L., Bernier, E. and Sullivan, M. J., 'An experimental investigation of the effect of a justice violation on pain experience and expression among individuals with high and low just world beliefs', *European Journal of Pain*, 18(3), 2014, pp.415–23

[36] Bissell, D. A., Ziadni, M. S. and Sturgeon, J. A., 'Perceived injustice in chronic pain: an examination through the lens of predictive processing', *Pain Management*, 8(2), 2018, pp.129–38

[37] Rodkey, E. N. and Riddell, R. P., 'The infancy of infant pain research: the experimental origins of infant pain denial', *Journal of Pain*, 14(4), 2013, pp.338–50

[38] Rovner S., 'Surgery without anesthesia: can preemies feel pain?', *Washington Post*, 13 August 1986

[39] Anand, K. J., Sippell, W. G. and Green, A. A., 'Randomised trial of fentanyl anaesthesia in preterm babies undergoing surgery: effects on the stress response', *The Lancet*, 329(8527), 1987, pp.243–8

[40] Raja, S. N., Carr, D. B., Cohen, M. *et al.*, 'The revised International Association for the Study of Pain definition of pain: concepts, challenges, and compromises', *Pain*, 161(9), 2020, pp.1976–82

[41] Goksan, S., Hartley, C., Emery, F. *et al.*, 'fMRI reveals neural activity overlap between adult and infant pain', *eLife*, 4, 2015, p.e06356

[42] Hartley, C., Goksan, S., Poorun, R. *et al.*, 'The relationship between nociceptive brain activity, spinal reflex withdrawal and behaviour in newborn infants', *Scientific Reports*, 5, 2015, p.12519

[43] Williams, M. D. and Lascelles, B. D. X., 'Early neonatal pain – review of clinical and experimental implications on painful conditions later in life', *Frontiers in Pediatrics*, 8, 2020

[44] van den Bosch, G. E., White, T., El Marroun, H. *et al.*, 'Prematurity, opioid exposure and neonatal pain: do they affect the developing brain?', *Neonatology*, 108(1), 2015, pp.8–15

[45] Hartley, C., Duff, E. P., Green, G. *et al.*, 'Nociceptive brain activity as a measure of analgesic efficacy in infants', *Science Translational Medicine*, 9(388), 2017, p.eaah6122

[46] Hartley, C., Moultrie, F., Hoskin, A. *et al.*, 'Analgesic efficacy and safety of morphine in the Procedural Pain in Premature Infants (Poppi) study: randomised placebo-controlled trial', *The Lancet*, 392(10164), 2018, pp.2595–605

[47] Brauer, J., Xiao, Y., Poulain, T., Friederici, A. D. and Schirmer, A., 'Frequency of maternal touch predicts resting activity and connectivity of the developing social brain', *Cerebral Cortex*, 26(8), 2016, pp.3544–52

[48] Liljencrantz, J. and Olausson, H., 'Tactile C fibers and their contributions to pleasant sensations and to tactile allodynia', *Frontiers in Behavioral Neuroscience*, 8, 2014

[49] Liljencrantz, J., Strigo, I., Ellingsen, D. M. *et al.*, 'Slow brushing reduces heat pain in humans', *European Journal of Pain*, 21(7), 2017, pp.1173–85

[50] Gursul, D., Goksan, S., Hartley, C. *et al*, 'Stroking modulates noxious-evoked brain activity in human infants', *Current Biology*, 28(24), 2018, pp.R1380–1

9 Belief as Relief

[1] Clark, W. C. and Clark, S. B., 'Pain responses in Nepalese porters', *Science*, 209(4454), 1980, pp.410–12

[2] Sargent, C. F., '*Maternity, Medicine, and Power: Reproductive Decisions in Urban Benin*', University of California Press, 1989

[3] Sternbach, R. A. and Tursky, B., 'Ethnic differences among housewives in psychophysical and skin potential responses to electric shock', *Psychophysiology*, 1(3), 1965, pp.241–6

[4] Kim, H. J., Yang, G. S., Greenspan, J. D. *et al.*, 'Racial and ethnic differences in experimental pain sensitivity: systematic review and meta-analysis', *Pain*, 158(2), 2017, pp.194–211

[5] Nayak, S., Shiflett, S. C., Eshun, S. and Levine, F. M., 'Culture and gender effects in pain beliefs and the prediction of pain tolerance', *Cross-Cultural Research*, 34(2), 2000, pp.135–51

[6] Dragioti, E., Tsamakis, K., Larsson, B. and Gerdle, B., 'Predictive association between immigration status and chronic pain in the general population: results from the SwePain cohort', *BMC Public Health*, 20(1), 2020, pp.1–11

[7] Kim, H. J., Greenspan, J. D., Ohrbach, R. *et al.*, 'Racial/ethnic differences in experimental pain sensitivity and associated factors – cardiovascular responsiveness and psychological status', *PLOS ONE*, 14(4), 2019, p.e0215534

[8] Byrne, M., Callahan, B., Carlson, K. *et al.*, *Nursing: A Concept-Based Approach to Learning*, ed. Trakalo, K., vol. 1., 2014

[9] Wiech, K., Farias, M., Kahane, G., Shackel, N., Tiede, W. and Tracey, I., 'An fMRI study measuring analgesia enhanced by religion as a belief system', *Pain*, 139(2), 2008, pp.467–76

[10] Ferreira-Valente, A., Sharma, S., Torres, S. *et al.*, 'Does religiosity/spirituality play a role in function, pain-related beliefs, and coping in patients with chronic pain? A systematic review', *Journal of Religion and Health*, 2019, pp.1–55

[11] Marx, K., *Critique of Hegel's 'Philosophy of Right'*, ed. O'Malley, J., Cambridge University Press, 2009

[12] Brand, P. and Yancey, P., *Pain: The Gift Nobody Wants*, HarperCollins, 1995

[13] Al-Bukhari, M., *Sahih al-Bukhari*, Mohee Uddin, 2020

[14] Alembizar, F., Hosseinkhani, A. and Salehi, A., 'Anesthesia and pain relief in the history of Islamic medicine', *Iranian Journal of Medical Sciences*, 41(3 Suppl), 2016, p.S21

[15] Sallatha, S., 'The Arrow', trans. Bhikkhu, T., *Access to Insight*, 1997

[16] 1 Peter 4:13, *The Bible* (English Standard Version)

[17] Revelation 21:4, *The Bible* (English Standard Version)

[18] Chou, R., Qaseem, A., Snow, V. *et al.*, 'Diagnosis and treatment of low back pain: a joint clinical practice guideline from the American College of Physicians and the American Pain Society', *Annals of Internal Medicine*, 147(7), 2007, pp.478–91

[19] Brinjikji, W., Luetmer, P. H., Comstock, B. *et al.*, 'Systematic literature review of imaging features of spinal degeneration in asymptomatic populations', *American Journal of Neuroradiology*, 36(4), 2015, pp.811–16

[20] Vibe Fersum, K., O'Sullivan, P., Skouen, J. S., Smith, A. and Kvåle, A., 'Efficacy of classification-based cognitive functional therapy in patients

with non-specific chronic low back pain: a randomized controlled trial', *European Journal of Pain*, 17(6), 2013, pp.916–28

[21] Vibe Fersum, K., Smith, A., Kvåle, A., Skouen, J. S. and O'Sullivan, P., 'Cognitive functional therapy in patients with non-specific chronic low back pain – a randomized controlled trial 3-year follow-up', *European Journal of Pain*, 23(8), 2019, pp.1416–24

10 The Silent Pandemic

[1] Fayaz, A., Croft, P., Langford, R. M., Donaldson, L. J. and Jones, G. T., 'Prevalence of chronic pain in the UK: a systematic review and meta-analysis of population studies', *BMJ Open*, 6(6), 2016, p.e010364

[2] Shipton, E. E., Bate, F., Garrick, R., Steketee, C., Shipton, E. A. and Visser, E. J., 'Systematic review of pain medicine content, teaching, and assessment in medical school curricula internationally', *Pain and Therapy*, 7(2), 2018, pp.139–61

[3] Blyth, F. M., March, L. M., Brnabic, A. J., Jorm, L. R., Williamson, M. and Cousins, M. J., 'Chronic pain in Australia: a prevalence study', *Pain*, 89(2–3), 2001, pp.127–34

[4] Sá, K. N., Moreira, L., Baptista, A. F. *et al.*, 'Prevalence of chronic pain in developing countries: systematic review and meta-analysis', *Pain Reports*, 4(6), 2019, p.e779

[5] McQuay, H., 'Help and hope at the bottom of the pile', *BMJ*, 336(7650), 2008, pp.954–5

[6] Treede, R. D., Rief, W., Barke, A. *et al.*, 'Chronic pain as a symptom or a disease: the IASP Classification of Chronic Pain for the International Classification of Diseases (ICD-11)', *Pain*, 160(1), 2019, pp.19–27

[7] Dyer, O., 'US life expectancy falls for third year in a row', *BMJ*, 363, 2018

[8] 'Odds of dying', *Injury Facts*, https://injuryfacts.nsc.org

[9] Olfson, M., Wall, M., Wang, S., Crystal, S. and Blanco, C., 'Service use preceding opioid-related fatality', *American Journal of Psychiatry*, 175(6), 2018, pp.538–44

[10] Krebs, E. E., Gravely, A., Nugent, S. *et al.*, 'Effect of opioid vs nonopioid medications on pain-related function in patients with chronic back pain or

hip or knee osteoarthritis pain: the SPACE randomized clinical trial', *JAMA*, 319(9), 2018, pp.872–82

[11] King, A., 'Analgesia without opioids', *Nature*, 573(7773), 2019, pp.S4-S6

[12] Rivat, C. and Ballantyne, J., 'The dark side of opioids in pain management: basic science explains clinical observation', *Pain Reports*, 1(2), 2016, p.e570

[13] Colvin, L. A., Bull, F. and Hales, T. G., 'Perioperative opioid analgesia – when is enough too much? A review of opioid-induced tolerance and hyperalgesia', *The Lancet*, 393(10180), 2019, pp.1558–68

[14] 'Opioids aware', Faculty of Pain Medicine, https://fpm.ac.uk/opioids-aware

[15] Pavlovic, S., Daniltchenko, M., Tobin, D. J. *et al.*, 'Further exploring the brain–skin connection: stress worsens dermatitis via substance P-dependent neurogenic inflammation in mice', *Journal of Investigative Dermatology*, 128(2), 2008, pp.434–46

[16] Liu, Y., Zhou, L. J., Wang, J. *et al.*, 'TNF-α differentially regulates synaptic plasticity in the hippocampus and spinal cord by microglia-dependent mechanisms after peripheral nerve injury', *Journal of Neuroscience*, 37(4), 2017, pp.871–81

[17] Hayley, S., 'The neuroimmune-neuroplasticity interface and brain pathology', *Frontiers in Cellular Neuroscience*, 8, 2014, p.419

[18] Araldi, D., Bogen, O., Green, P. G. and Levine, J. D., 'Role of nociceptor Toll-like Receptor 4 (TLR4) in opioid-induced hyperalgesia and hyperalgesic priming', *Journal of Neuroscience*, 39(33), 2019, pp.6414–24

[19] Evers, A. W. M., Verhoeven, E. W. M., Kraaimaat, F. W. *et al.*, 'How stress gets under the skin: cortisol and stress reactivity in psoriasis', *British Journal of Dermatology*, 163(5), 2010, pp.986–91

[20] Young, M. B., Howell, L. L., Hopkins, L. *et al.*, 'A peripheral immune response to remembering trauma contributes to the maintenance of fear memory in mice', *Psychoneuroendocrinology*, 94, 2018, pp.143–51

[21] Goshen, I., Kreisel, T., Ounallah-Saad, H. *et al.*, 'A dual role for interleukin-1 in hippocampal-dependent memory processes', *Psychoneuroendocrinology*, 32(8–10), 2007, pp.1106–15

[22] Michopoulos, V., Powers, A., Gillespie, C. F., Ressler, K. J. and Jovanovic, T., 'Inflammation in fear- and anxiety-based disorders: PTSD, GAD, and beyond', *Neuropsychopharmacology*, 42(1), 2017, pp.254–70

[23] Burke, N. N., Finn, D. P., McGuire, B. E. and Roche, M., 'Psychological stress in early life as a predisposing factor for the development of chronic pain: clinical and preclinical evidence and neurobiological mechanisms', *Journal of Neuroscience Research*, 95(6), 2017, pp.1257–70

[24] Bower, J .E. and Irwin, M. R., 'Mind–body therapies and control of inflammatory biology: a descriptive review', *Brain, Behavior, and Immunity*, 51, 2016, pp.1–11

[25] Smith, K., 'The association between loneliness, social isolation and inflammation: a systematic review and meta-analysis', *Neuroscience & Biobehavioral Reviews*, 112, 2020, pp.519–41

[26] Hussain, S. M., Urquhart, D. M., Wang, Y. *et al.*, 'Fat mass and fat distribution are associated with low back pain intensity and disability: results from a cohort study', *Arthritis Research & Therapy*, 19, 2017, p.26

[27] Smuck, M., Schneider, B. J., Ehsanian, R., Martin, E. and Kao, M. C. J., 'Smoking is associated with pain in all body regions, with greatest influence on spinal pain', *Pain Medicine*, 21(9), 2020, pp.1759–68

[28] Morin, C. M., LeBlanc, M., Daley, M., Gregoire, J. P. and Merette, C., 'Epidemiology of insomnia: prevalence, self-help treatments, consultations, and determinants of help-seeking behaviors', *Sleep Medicine*, 7(2), 2006, pp.123–30

[29] Taylor, D. J., Mallory, L. J., Lichstein, K. L., Durrence, H. H., Riedel, B. W. and Bush, A. J., 'Comorbidity of chronic insomnia with medical problems', *Sleep*, 30(2), 2007, pp.213–18

[30] Gerhart, J. I., Burns, J. W., Post, K. M. *et al.*, 'Relationships between sleep quality and pain-related factors for people with chronic low back pain: tests of reciprocal and time of day effects', *Annals of Behavioral Medicine*, 51(3), 2017, pp.365–75

[31] Krause, A. J., Prather, A. A., Wager, T. D., Lindquist, M. A. and Walker, M. P., 'The pain of sleep loss: a brain characterization in humans', *Journal of Neuroscience*, 39(12), 2019, pp.2291–300

[32] Irwin, M. R., Wang, M., Ribeiro, D. *et al.*, 'Sleep loss activates cellular inflammatory signaling', *Biological Psychiatry*, 64(6), 2008, pp.538–40

[33] Billari, F. C., Giuntella, O. and Stella, L., 'Broadband internet, digital temptations, and sleep', *Journal of Economic Behavior & Organization*, 153, 2018, pp.58–76

[34] Lam, K. K., Kunder, S., Wong, J., Doufas, A. G. and Chung, F., 'Obstructive sleep apnea, pain, and opioids: is the riddle solved?', *Current Opinion in Anaesthesiology*, 29(1), 2016, pp.134–40

[35] Moore, J. T. and Kelz, M. B., 'Opiates, sleep, and pain: the adenosinergic link', *Anesthesiology*, 111(6), 2009, pp.1175–76

11 The Runaway Brain

[1] Woolf C. J., 'Evidence for a central component of post-injury pain hypersensitivity', *Nature*, 306, 1983, pp.686–8

[2] Sandkühler, J. and Gruber-Schoffnegger, D., 'Hyperalgesia by synaptic long-term potentiation (LTP): an update', *Current Opinion in Pharmacology*, 12(1), 2012, pp.18–27

[3] Jepma, M., Koban, L., van Doorn, J., Jones, M. and Wager, T.D., 'Behavioural and neural evidence for self-reinforcing expectancy effects on pain', *Nature Human Behaviour*, 2(11), 2018, pp.838–55

[4] Soni, A., Wanigasekera, V., Mezue, M. *et al.*, 'Central sensitization in knee osteoarthritis: relating presurgical brainstem neuroimaging and PainDETECT-based patient stratification to arthroplasty outcome', *Arthritis & Rheumatology*, 71(4), 2019, pp.550–60

[5] Tagliazucchi, E., Balenzuela, P., Fraiman, D. and Chialvo, D. R., 'Brain resting state is disrupted in chronic back pain patients', *Neuroscience Letters*, 485(1), pp.26–31

[6] Apkarian, A. V., Sosa, Y., Sonty, S. *et al.*, 'Chronic back pain is associated with decreased prefrontal and thalamic gray matter density', *Journal of Neuroscience*, 24(46), 2004, pp.10410–15

[7] Johnston, K. J., Adams, M. J., Nicholl, B. I. *et al.*, 'Genome-wide association study of multisite chronic pain in UK Biobank', *PLOS Genetics*, 15(6), 2019, p.e1008164

[8] Khoury, S., Piltonen, M. H., Ton, A. T. *et al.*, 'A functional substitution in the L-aromatic amino acid decarboxylase enzyme worsens somatic symptoms via a serotonergic pathway', *Annals of Neurology*, 86(2), 2019, pp.168–80

[9] Desmeules, J. A., Cedraschi, C., Rapiti, E. *et al.*, 'Neurophysiologic evidence for a central sensitization in patients with fibromyalgia', *Arthritis & Rheumatism*, 48(5), 2003, pp.1420–9

[10] Cagnie, B., Coppieters, I., Denecker, S., Six, J., Danneels, L. and Meeus, M., 'Central sensitization in fibromyalgia? A systematic review on structural and functional brain MRI', *Seminars in Arthritis and Rheumatism*, 44(1), 2014, pp.68–75

[11] Bäckryd, E., Tanum, L., Lind, A. L., Larsson, A. and Gordh, T., 'Evidence of both systemic inflammation and neuroinflammation in fibromyalgia patients, as assessed by a multiplex protein panel applied to the cerebrospinal fluid and to plasma', *Journal of Pain Research*, 10, 2017, pp.515–25

[12] Albrecht, D. S., Forsberg, A., Sandström, A. *et al.*, 'Brain glial activation in fibromyalgia – a multi-site positron emission tomography investigation', *Brain, Behavior, and Immunity*, 75, 2019, pp.72–83

[13] Stankevicius, A., Wallwork, S. B., Summers, S. J., Hordacre, B. and Stanton, T. R., 'Prevalence and incidence of phantom limb pain, phantom limb sensations and telescoping in amputees: a systematic rapid review', *European Journal of Pain*, 25(2), 2020

[14] Weinstein, S. M., 'Phantom limb pain and related disorders', *Neurologic Clinics*, 16(4), 1998, pp.919–35

[15] Penfield, W. and Jasper, H., *Epilepsy and the Functional Anatomy of the Human Brain*, Little, Brown, 1954

[16] Ramachandran, V. S., 'Perceptual Correlates of Neural Plasticity in the Adult Human Brain', *Early Vision and Beyond*, eds. Papathomas, T. V., Kowler, E., Chubb, C. and Gorea, A., MIT Press, 1995, pp.227–47

[17] Flor, H., Nikolajsen, L. and Jensen, T. S., 'Phantom limb pain: a case of maladaptive CNS plasticity?', *Nature Reviews Neuroscience*, 7(11), 2006, pp.873–81

[18] Flor, H., Elbert, T., Knecht, S. *et al.*, 'Phantom-limb pain as a perceptual correlate of cortical reorganization following arm amputation', *Nature*, 375(6531), pp.482–4

[19] Ramachandran, V. S. and Blakeslee, S., *Phantoms in the Brain*, Fourth Estate, 1999

[20] Doidge, N., *The Brain That Changes Itself: Stories of Personal Triumph from the Frontiers of Brain Science*, Penguin, 2008

[21] Freeman, M. D., Nystrom, A. and Centeno, C., 'Chronic whiplash and central sensitization; an evaluation of the role of a myofascial trigger point in pain modulation', *Journal of Brachial Plexus and Peripheral Nerve Injury*, 4(1), 2009, pp.1–8

[22] Campo-Prieto, P. and Rodríguez-Fuentes, G., 'Effectiveness of mirror therapy in phantom limb pain: a literature review', *Neurología*, English edition, 2018

[23] McCabe, C. S., Haigh, R. C., Ring, E. F. J., Halligan, P. W., Wall, P. D. and Blake, D. R., 'A controlled pilot study of the utility of mirror visual feedback in the treatment of complex regional pain syndrome (type 1)', *Rheumatology*, 42(1), 2003, pp.97–101

[24] Bowering, K. J., O'Connell, N. E., Tabor, A. *et al.*,'The effects of graded motor imagery and its components on chronic pain: a systematic review and meta-analysis', *Journal of Pain*, 14(1), 2013, pp.3–13

[25] Kikkert, S., Mezue, M., O'Shea, J. *et al.*, 'Neural basis of induced phantom limb pain relief', *Annals of Neurology*, 85(1), 2019, pp.59–73

[26] Rutledge, T., Velez, D., Depp, C. *et al.*, 'A virtual reality intervention for the treatment of phantom limb pain: development and feasibility results', *Pain Medicine*, 20(10), 2019, pp.2051–9

12 The Pain Revolution

[1] Corkhill, B., *Knitting for Health and Wellness*, Flatbear Publishing, 2014

[2] Riley, J., Corkhill, B. and Morris, C., 'The benefits of knitting for personal and social wellbeing in adulthood: findings from an international survey', *British Journal of Occupational Therapy*, 76(2), 2013, pp.50–7

[3] Jacobs, B. L. and Fornal, C. A., 'Activity of serotonergic neurons in behaving animals', *Neuropsychopharmacology*, 21(1), 1999, pp.9–15

[4] Draganski, B., Gaser, C., Busch, V., Schuierer, G., Bogdahn, U. and May, A., 'Changes in grey matter induced by training', *Nature*, 427(6972), 2004, pp.311–12

[5] Gallace, A., Torta, D. M. E., Moseley, G. L. and Iannetti, G. D., 'The analgesic effect of crossing the arms', *Pain*, 152(6), 2011, pp.1418–23

[6] McKay, J. H. and Tatum, W. O., 'Knitting induced fronto-central theta rhythm', *Epilepsy & Behavior Reports*, 12, 2019, p.100335

[7] Corkhill, B. and Davidson, C., 'Exploring the effects of knitting on the experience of chronic pain – a qualitative study', poster at the British Pain Society Annual Scientific Meeting, 2009

[8] Ponce-Alonso, M., de la Fuente, J. S., Rincón-Carlavilla, A. *et al.*, 'Impact of the coronavirus disease 2019 (COVID-19) pandemic on nosocomial *Clostridioides difficile* infection', *Infection Control & Hospital Epidemiology*, 2020, pp.1–5

[9] Greenhalgh, T., 'Pondering whether COVID-19 will be evidence-based medicine's nemesis', Twitter post, 2 May 2020

[10] Tremblay, M. S., Colley, R. C., Saunders, T. J., Healy, G. N. and Owen, N., 'Physiological and health implications of a sedentary lifestyle', *Applied Physiology, Nutrition, and Metabolism*, 35(6), 2010, pp.725–40

[11] Hanna, F., Daas, R. N., El-Shareif, T. J., Al-Marridi, H. H., Al-Rojoub, Z. M. and Adegboye, O. A., 'The relationship between sedentary behavior, back pain, and psychosocial correlates among university employees', *Frontiers in Public Health*, 7, 2019, p.80

[12] Heron, L., O'Neill, C., McAneney, H., Kee, F. and Tully, M. A., 'Direct healthcare costs of sedentary behaviour in the UK', *Journal of Epidemiolgy and Community Health*, 73(7), 2019, pp.625–9

[13] Gopinath, B., Kifley, A., Flood, V. M. and Mitchell, P., 'Physical activity as a determinant of successful aging over ten years', *Scientific Reports*, 8(1), 2018, pp.1–5

[14] Rice, D., Nijs, J., Kosek, E. *et al.*, 'Exercise-induced hypoalgesia in pain-free and chronic pain populations: state of the art and future directions', *Journal of Pain*, 20(11), 2019, pp.1249–66

[15] Dimitrov, S., Hulteng, E. and Hong, S., 'Inflammation and exercise: inhibition of monocytic intracellular TNF production by acute exercise via β2-adrenergic activation', *Brain, Behavior, and Immunity*, 61, 2017, pp.60–8

[16] Puetz, T. W., Flowers, S. S. and O'Connor, P. J., 'A randomized controlled trial of the effect of aerobic exercise training on feelings of energy and fatigue in sedentary young adults with persistent fatigue', *Psychotherapy and Psychosomatics*, 77(3), 2008, pp.167–74

[17] Nijs, J., Girbés, E. L., Lundberg, M., Malfliet, A. and Sterling, M., 'Exercise therapy for chronic musculoskeletal pain: innovation by altering pain memories', *Manual Therapy*, 20(1), 2015, pp.216–20

[18] 'The Health and Wellbeing Benefits of Swimming', Swimming and Health Commission, 2017

[19] Busch, V., Magerl, W., Kern, U., Haas, J., Hajak, G. and Eichhammer, P., 'The effect of deep and slow breathing on pain perception, autonomic activity, and mood processing – an experimental study', *Pain Medicine*, 13(2), 2012, pp.215–28

[20] Anderson, B. E. and Bliven, K. C. H., 'The use of breathing exercises in the treatment of chronic, nonspecific low back pain', *Journal of Sport Rehabilitation*, 26(5), 2017, pp.452–8

[21] Gerhart, J. I., Burns, J. W., Post, K. M. *et al.*, 'Relationships between sleep quality and pain-related factors for people with chronic low back pain: tests of reciprocal and time of day effects', *Annals of Behavioral Medicine*, 51(3), 2017, pp.365–75

[22] Brasure, M., Fuchs, E., MacDonald, R. *et al.*, 'Psychological and behavioral interventions for managing insomnia disorder: an evidence report for a clinical practice guideline by the American College of Physicians', *Annals of Internal Medicine*, 165(2), 2016, pp.113–24

[23] Finan, P. H., Buenaver, L. F., Runko, V. T. and Smith, M. T., 'Cognitive-behavioral therapy for comorbid insomnia and chronic pain', *Sleep Medicine Clinics*, 9(2), 2014, pp.261–74

[24] Sapolsky, R. M., *Why Zebras Don't Get Ulcers: The Acclaimed Guide to Stress, Stress-related Diseases, and Coping*, Holt, 2004

[25] Doidge, N., *The Brain's Way of Healing: Remarkable Discoveries and Recoveries from the Frontiers of Neuroplasticity*, Penguin, 2016

[26] Moseley, G. L., Parsons, T. J. and Spence, C., 'Visual distortion of a limb modulates the pain and swelling evoked by movement', *Current Biology*, 18(22), 2008, pp.R1047–8

[27] Stanton, T. R., Gilpin, H. R., Edwards, L., Moseley, G. L. and Newport, R., 'Illusory resizing of the painful knee is analgesic in symptomatic knee osteoarthritis', *PeerJ*, 6, 2018, p.e5206

[28] Butler, D. S. and Moseley, G. L., *Explain Pain*, 2nd edition, NOI Group, 2013

[29] Moseley, G. L., 'Evidence for a direct relationship between cognitive and physical change during an education intervention in people with chronic low back pain', *European Journal of Pain*, 8(1), 2004, pp.39–45

[30] Moseley, G. L. and Butler, D. S., 'Fifteen years of explaining pain: the past, present, and future', *Journal of Pain*, 16(9), 2015, pp.807–13

[31] Louw, A., Zimney, K., Puentedura, E. J. and Diener, I., 'The efficacy of pain neuroscience education on musculoskeletal pain: a systematic review of the literature', *Physiotherapy Theory and Practice*, 32(5), 2016, pp.332–55

[32] Lee, H., McAuley, J. H., Hübscher, M., Kamper, S. J., Traeger, A. C. and Moseley, G. L., 'Does changing pain-related knowledge reduce pain and improve function through changes in catastrophizing?', *Pain*, 157(4), 2016, pp.922–30

[33] Corrigan, C., Desnick, L., Marshall, S., Bentov, N. and Rosenblatt, R. A., 'What can we learn from first-year medical students' perceptions of pain in the primary care setting?', *Pain Medicine*, 12(8), 2011, pp.1216–22

[34] Mackey, C., 'Pain and the Brain', lecture at Stanford Back Pain Education Day 2016, Youtube.com

Acknowledgements

The Painful Truth is for everyone, but I sincerely hope that it is helpful for, and does justice to, those living with persistent (chronic) pain. Some of these people have been dismissed by medical professionals; many have been told that their pain is all in their head; all have suffered. If this book does anything, I hope that it raises awareness of the seriousness of persistent pain at the individual and societal level. Without meeting these people as either their doctor or interviewer, this book would not have been possible. Only a few – such as Evan – have I been able to name; the majority remain behind pseudonyms. Thank you.

The second group to whom I am indebted are the scientists and clinicians who have dedicated their lives to understanding and relieving pain. *The Painful Truth* contains almost four hundred references – representing over one thousand researchers – but these are only a tiny fraction of the remarkable people who have dared to think differently, giving us a modern understanding of pain.

To my utterly brilliant editor, Andrea Henry, and the rest of the team at Transworld: Tom Hill, Kate Samano, Phil Lord, Alex Newby and Richard Shailer.

To my agent, Charlie Viney, for his foresight in predicting that pain would make a fascinating topic.

To my supervisors and mentors, who have inspired my love of writing, science and the human condition: Kate Thomas, Colin Thubron, Margreta de Grazia, Dafydd Lloyd, Daqing Ma, Graham Ogg, Olga Tsatalou, Kate Dean, Belinda Lennox and John Beale.

To those who kindly gave their time and experience to be interviewed for this book: Jo Cameron, Candice, Betsan Corkhill, Deepak Ravindran, Joel Salinas, James Robinson, Paul Hughes, Denise Gursul, Tim Keller, Ithsham Iqbal and, of course, those who can't be named.

To my parents, Rob and Hannah, whose example inspired me to write and pursue a caring profession. And to my brother, Phin, for providing me with pain-related material in the way that only younger brothers can.

To Hannah, my wife. You are a sage and a sounding board, and I could not have written this without you.

And, finally, to you, the reader. Thank you for reading my book. I hope that it at least broadens, if not completely changes your understanding of pain. Remain curious, keep learning and spread the word. Understanding pain is the key to relieving it.

Index

INDEX

INDEX

INDEX

INDEX

THE REMARKABLE LIFE OF THE SKIN
An intimate journey across our surface

Monty Lyman

The skin is our largest and fastest-growing organ. We see it, touch it and live in it every day. It's a habitat for a mesmerizingly complex world of micro-organisms and its physical functions are vital to our health and indeed our survival. Yet how much do we really know about the skin? What makes it age? How does our diet affect it? Why can't we tickle ourselves?

Through the lenses of science, sociology and history, Dr Monty Lyman reveals how the skin is far stranger and more complex than you've ever imagined.

'A seriously entertaining book.' *The Times*

'An exciting introduction to a little-known microscopic universe and to a talented new writer.' *Sunday Times*

'Lyman peels back the science on human skin in this absorbing, fact-packed study.' *Nature*

'Imaginatively told . . . knowledgeable and enthusiastic.' *Spectator*